THE CRESTLINE SERIES
GAS STATIONS

Wayne Henderson
Scott Benjamin

First published in 1994 by Motorbooks International Publishers & Wholesalers, PO Box 2, 729 Prospect Avenue, Osceola, WI 54020 USA

© Wayne Henderson, 1994

All rights reserved. With the exception of quoting brief passages for the purpose of review no part of this publication may be reproduced without prior written permission from the Publisher

Motorbooks International is a certified trademark, registered with the United States Patent Office

The information in this book is true and complete to the best of our knowledge. All recommendations are made without any guarantee on the part of the author or Publisher, who also disclaim any liability incurred in connection with the use of this data or specific details

We recognize that some words, model names and designations, for example, mentioned herein are the property of the trademark holder. We use them for identification purposes only. This is not an official publication

Motorbooks International books are also available at discounts in bulk quantity for industrial or sales-promotional use. For details write to Special Sales Manager at the Publisher's address

Library of Congress Cataloging-in-Publication Data

Henderson, Wayne.
  Gas stations/Wayne Henderson and Scott Benjamin.
     p. cm.
  Includes index.
  ISBN 0-87938-945-1
  1. Service stations—United States—History. I. Benjamin, Scott. II. Title.
  TL153.H46   1994
  629.28'6'0973—dc20   94-23073

On the front cover: When Walt Wimer, Jr., pulled into this Milwaukee, Wisconsin, Pate station in 1959, he was greeted by the Pate sign's warm "Hello Neighbor!" welcome. Walt also visited the classic Phillips cottage station near Oklahoma City, Oklahoma. *Walt Wimer, Jr.*

On the back cover: In 1958, Walt Wimer, Jr., started a long road trip in his 1957 Chevrolet. He photographed many gas stations along the way, including this Vickers station in remote Rexford, Kansas. Walt also visited an Imperial station in DeLand, Florida, in 1960, when an island kiosk not unlike those found at today's self-serve stations was also in use. *Walt Wimer, Jr.*

Printed and bound in the United States of America

Contents

Introduction	**Corner Store & Country Store... To Convenience Store** *Gasoline Marketing Comes Full Circle*	4
Chapter 1	**1910–1918: The Early Years—Curb Pumps and Country Stores**	10
Chapter 2	**1918-1929** *The Filling Station—Standardized Designs and The Two-Post Canopy*	13
Chapter 3	**1929-1942** *The Depression—Gas Stations Come Of Age*	35
Chapter 4	**1942-1945** *The War Years—Rationing and Restrictions*	88
Chapter 5	**1945-1954** *Early Postwar Years*	91
Chapter 6	**1954-1965** *The Creation of the Miracle Mile*	119
Chapter 7	**1965-1974** *Interstates and 'Keep America Beautiful'*	185
Chapter 8	**1974-1985** *Shortages, Self-Serve, and C-Stores*	195
Chapter 9	**1985-1994** *Modern Times—'Images and Environmental Concerns'*	204
Chapter 10	**Survivors**	211
	Index	224

Introduction

Corner Store & Country Store... To Convenience Store

Gasoline Marketing Comes Full Circle

The work day is over and the motorist hurries home at dusk. A quick stop for gasoline is necessary. He eases his car as close to the gas pump as possible and stops the engine. Getting out of the auto, he notices the improved lighting at the station.

"About time," he thinks. "How do they expect to attract customers without first-class lighting?"

He quickly fills the tank and enters the store to pay. "Might as well pick up a few things since I'm here." Eyeing the menu board behind the counter, he decides a sandwich would spoil supper, but a fountain drink might just hit the spot. Selecting a headache remedy from one of several available, and a pack of cigarettes, he heads to the counter to check out. He pays for his purchases and heads for his automobile to drive off into the night.

A typical scene in any city or town in America, right? How about if the improved lighting was the addition of a one-piece etched "filtered gasoline" globe to the top of the hand-operated curb pump? What if the headache remedy is easily selected because the establishment is a corner drug store? What if his purchases consist of a fountain "Moxie" and a pack of "Piedmont" cigarettes? And what if the automobile that carries him home is his new 1911 Ford Model T?

The single most familiar sight to ever appear on the commercial American landscape is the gas station. Whether it's a curb pump in front of a grocery store on main street, a tall visible pump beside a kerosene tank at a country store, a brick "filling station" with two posts supporting an attached canopy, a trackside discounter on the gravel lot down by the railroad tracks, a two-bay dealer-operated service station constructed of porcelain panels or in rancher style brick, a self-serve kiosk under an expansive canopy, or an ultramodern travel plaza with a convenience store and fast food franchise all under one roof, the image is still simply "gas station," the most repeated commercial institution of this century.

Let us take a look back, and maybe even something of a look ahead, at this roadside institution.

The Earliest Stations

Collectors of gasoline memorabilia are often provided information on types of collectibles, market trends and events, oil companies and their histories, but most likely they've read very little on where the institution of the roadside service station came from and how it evolved. A quick look at the development of gasoline marketing will not only provide an interesting history lesson, but also will reveal new places to look for gasoline memorabilia and insight into why items from some companies are much harder to find than others.

Turn-of-the-century motorists, few in number, searched for gasoline almost as an industrial commodity. Hardware stores were a prime source, as were bulk lubricant and fuel storage plants, forerunners of today's gasoline jobbers. A motorist would purchase gasoline in 1 gallon (gal) prepackaged containers or from bulk storage. Gasoline would be transferred from bulk into 5gal cans from which it would be funneled into the automobile gas tank. Credit for the first location in America where a motorist could have gasoline dispensed directly from storage into his car goes to Standard Oil of California—Chevron. In 1907 their bulk station operator in Seattle, Washington, devised a dispensing device made from an old water heater mounted on a stand like a fuel oil tank. Crude, but effective.

By 1910 the principles of modern underground storage with pump and hose dispensing had been developed, and curb pumps appeared on the streets of every city and town in the country. Hardware stores, feed and farm supply stores, and drug stores in cities and small towns quickly added curb pumps. And with these devices they added to the congestion of city streets already shared by automobiles, horse-drawn vehicles and streetcars. Out in the country, mercantile stores, having supplied everything area residents needed since the stores' development in the post-Civil War era, added a small storage tank and curb pump.

While the curb pump was ideal for the country store, with its large lot and often multiple buildings for different purposes, the congestion associated with sidewalk-mounted curb pumps in cities led to a new development. In December 1913, Gulf Oil Company opened the first gasoline filling station built for that purpose with off-street fueling. Located at the corner of Baum Boulevard and St. Clair Avenue in Pittsburgh, Pennsylvania, that station was the first true forerunner of the roadside gasoline station that would supply our automotive needs for the rest of this century.

By 1915 the concept of branded gasoline was being brought forth. Stations were labeled with new identification signs: "The Texas Company Filling Station," "Good Gulf Gasoline," and so forth. Advertising was devoted to the concept of differentiating between various gasoline brands. Forgotten was the fact that the motorist of just a few years earlier was glad to purchase *any* brand or quality of gasoline *anywhere* he could find it. Major oil companies, led by the newly freed Standard affiliates (Standard Oil was broken up in 1911), competed for sites to develop chains of company operated gasoline stations. Staffed by uniformed attendants, these earliest stations—oil company owned and operated by their own employees—sold only gasoline and lubricants. Elaborate station structures, from Atlantic's "palaces" in Pennsylvania to Wadhams' "pagodas" throughout Wisconsin, added respectability to oil industry marketing.

During this era, refining capacity finally caught up with and then surpassed demand. Refiners began to offer surplus unbranded gasoline on the open market and many former coal dealers, hardware store owners, or blacksmiths constructed bulk plants and became oil jobbers. Early jobbers sold under their own brand names or joined associations such as IOMA (Red Hat) or Dixie Distributors, and they rarely expanded their sales areas beyond a region consisting of several counties. They would salary operate some locations and supply to independent merchants who became gasoline dealers. Many of these independent dealers had entered gasoline marketing from other automotive endeavors—automobile dealers or repair garages—and as such, continued with an extensive repair business. Major oil companies took notice, and by the mid-1920s they were offering automotive services, mostly lubrication, at company owned major brand stations. The decade from 1920–1930 saw the construction of hundreds of thousands of roadside "service stations," as they had became known, consisting of a mixture of: major brand, company operated locations; a small number of major brand, dealer-operated locations; some jobber-owned and -operated locations, a few of which were identified by a major brand or brands; and a large number of jobber-supplied, independently owned locations.

The First Convenience Stores

A seemingly unrelated event occurred in Dallas, Texas, in 1927. In those days before refrigeration equipment was common, ice was supplied as a public commodity, much like natural gas or telephone service. While most ice was delivered by horse and wagon, customers not living on an established route purchased ice at local "ice docks."

In 1927 an ice dock manager working for Dallas-based Southland Ice Company added a few grocery items—milk, bread, and soft drinks—to his stock at the request of his customers. The experiment was so successful that Southland quickly added grocery items to all of their ice docks, and they became "Totem Ice" stores, identified by the genuine Totem Pole out in front. Totem Ice became "Seven-Eleven" in 1946, and the convenience store was born. Of interest to collectors of gasoline memorabilia is the fact that in 1928, Totem teamed up with Transcontinental Oil to locate a number of Transcontinental's "Marathon" gasoline stations adjacent to Totem Ice stores, creating the first gasoline/convenience-store tie-in.

Gasoline marketing saw another new innovation in 1928. That was when John Mason Houghland leased land along a railroad spur line in the Nashville suburb of Old Hickory and contracted to have gasoline delivered directly to the site in tank car lots. While other independent dealers and jobbers sold gasoline that had passed through several hands prior to its final sale, Houghland purchased direct from refineries in lots that were large enough to eliminate the middlemen and thus lower the price. "Discount" gasoline began with

Houghland's small station, which he branded "Spur."

Many such stations quickly sprang up, and many market experts made their way to Nashville or other cities where Spur stations were located to learn more about their operation. Southern cities, with access to products from Gulf Coast refineries, soon saw brands such as Thoni's "Magic Benzol Gas Stations," Dixie Vim, Tankar, Trackside, and others appear. By the early thirties, the industrial northern cities also saw the earliest efforts at discount gasoline through chain operations such as Red Head "Fair Price" Stations, Gaseteria/Bonded, Oklahoma, Clark, Martin, and Hudson. Also of note were chain operations that sprang up in Minnesota and Wisconsin. Duluth, Minnesota's "Webb Cut Price" stations was an early northern discounter, along with St. Paul-based North Star.

Of particular note, however, was the founding in the small Wisconsin town of Barron of what has been one of the most innovative marketers of all time—indeed, the company that taught the majors how they would market gasoline in the nineties and beyond. In 1931 the Erickson family, involved in wholesale grocery sales in Barron, opened a small gasoline station they called Erickson's Quick Serv. We'll look again at Erickson's a little later in the story.

Dealer-Operated Stations

The early thirties saw the passage of chain store taxes in numerous states. Initially designed to protect local grocers from the influx of chains such as A & P, the tax laws were interpreted to include retail gasoline stations. When Iowa levied a tax on each location operated by a single firm, Standard Oil of Indiana, with over 1,000 locations in the state, responded by firing all of their station personnel—and giving each of them the option of leasing the station that they had been operating. While the leases were structured so Standard remained very much in control, the stations did indeed become independent businesses, and the major brand service station dealer came into being. While some independent businessmen had represented some major oil companies back in the earliest days of service stations, primarily in areas not economically serviceable by the majors or in areas where major oil companies were represented by independent jobbers, this was the first major conversion of stations from company operated to dealer operated.

Automotive repair at service stations, barely a footnote on the sales charts before 1925, was becoming a common and accepted aspect of the business by 1930. It was in 1930 that Colonial Beacon Oil Company, having been recently purchased by Standard Oil of New Jersey (Esso), introduced a line of tires and batteries for sale at their stations. This led to the formation, of the Atlas Supply Company. Atlas, jointly owned by the Standard Oil Companies of New Jersey, Kentucky, California, Ohio, and Indiana, offered a full line of Tires, Batteries, and Accessories (known as "TBA") for marketing through each of the affiliated companies. The affiliates were chosen based on the fact that they operated under the "Standard" brand in their respective areas, and with those "assigned" territories, figured that they would never compete among themselves. Other marketers scrambled to keep up, and most majors were offering a national brand of TBA by 1935. The gasoline station of the next four decades—major-branded, dealer-operated, selling gasoline, lubricants, and doing automotive repair—had came into being.

Gasoline shortages during World War II forced dealers to rely more on repair work for survival and indeed forced many trackside discounters out of business. While wholesale gas was readily available to discounters when times were good, the shortages during the war caused the majors to divert supplies to their own branded operations, leaving unbranded purchasers with no source of supply. Those with long-term supply contracts survived the war only to find that anyone could sell gasoline at *any* price in the postwar boom. The built-up demand for new cars, the factory payrolls, and the returning military personnel were all factors in the boom, and discounters lost the attractiveness of their price advantage.

Several discounters, notably California-based Urich Oil and Virginia-based Tankar Stations, introduced self-service at locations in the late forties to differentiate themselves from the majors, cut operating costs, and to further cut retail price. Payless, the New Albany, Indiana, discounter, as well as several other chains that operated in "big city" settings in the industrial northeast, invented the gasoline "pumper" in this era. These were large, multi-pump gasoline stations with a small central cashier's stand under a large canopy that covered most of the station lot. The improved lighting, impressive layout, and canopy covered islands (great for northern cities' winters) improved the overall image of discounters. Others would follow these trends.

Changes in the Fifties

The fifties saw the discounters either expand into other fields—exploration, refining, and other functions of major oil companies—or sell out to the majors. The majors, bankrolled by profits from the postwar boom, purchased dozens of independents and chains. Standard of New Jersey (Esso) led the way by purchasing hundreds of locations from the formerly independent Oklahoma, Perfect Power, Pate, Deem, and Gaseteria/Bonded chains, eventually converting many to their "Enco" brand introduced in 1960.

Other marketers, including Clark, Erickson's, and Hudson, expanded into refining, while others joined with a new mid-level oil company known as an "integrated independent." Integrated independents had many of the same operations as major oil companies but without the large (geographically) service areas and without having to maintain far-flung and well-controlled marketing efforts.

Kerr McGee, Fina, and Ashland acquired stations over larger and larger areas. Even the originator, Spur, joined with a new integrated independent, Murphy Oil, in 1960. As the operations sold out, most were converted to major oil style operations—dealer-operated, repair-oriented, "Class A" stations.

One of the discounters that *did* survive the fifties introduced nineties' gasoline marketing—back in 1959. That was the year that Erickson's, with a refinery and more than 100 stations in four states, launched Holiday Stationstores. With the slogan "America's Most Unusual Service Stations," Erickson's converted their small gas-and-oil-only operations into full-blown convenience stores—groceries, health and beauty aids, automotive products, even sporting goods all under one roof. With canted store windows floor to ceiling, catchy signage, and multiple gas pump islands, the new Holiday Stationstores were indeed the most unusual service station most folks had ever seen. To the branded Texaco dealer and his supplying jobber, to the new Murphy-Spur dealer, even to the Clark station managers and regional managers who visited the new convenience stores and walked away thinking "It'll never catch on," the concept of groceries, hardware, and drugstore items being sold in a service station seemed farfetched.

They had forgotten that gasoline was sold in just such a setting forty to fifty years earlier. And while credit for the first convenience store and even the first C-store/gasoline tie-in goes to Southland/Seven-Eleven, Erickson's Holiday Stationstores represented the first effort by the oil industry—remember, Erickson's was a refiner—to enter the convenience store market. All the rest of the industry's players would eventually follow.

Interstates Change the Rules

By the time the petroleum industry celebrated its first centennial in 1959, it had been several years since President Eisenhower signed into law the bill that created the Interstate Highway System. The new highway network would have a larger impact on gasoline marketing than any event up to that point—almost immediately causing the obsolescence of thousands of once-thriving gas stations, and demanding revisions to those that would survive. As the interstate system was constructed, those many miles of two-lane road that had been truck, tourist, and commuter routes were bypassed by highways that made provisions for automotive refueling only at selected locations like travel plazas or interchanges. Only those stations fortunate enough to have been located near an access point to the new Interstate continued operating without significant change. If there was a change, in fact, it was that these surviving stations with close proximity to on- and off-ramps saw a significant increase in their business. Rural highway locations closed in record numbers. Truck stops, an invention of the thirties, when roadside gas stations began to operate restaurants and sell diesel fuel on a 24-hour basis to appeal to truckers, expanded greatly in this era.

Pure Oil affiliate Hickock Oil had led the way, and the Pure brand appeared on more truck stops than others that experimented with them, notably Pan-Am and Skelly. Surviving rural highway locations were modified for better high-speed visibility—with better lighting, larger internally lit plastic signage, even the addition of huge "high rise" identification signs—if they were located near an Interstate. All of this was designed to attract the attention of motorists passing at speeds requiring much more decision time before they could make a roadside stop.

Identification signs had evolved from lithographed tin or porcelain-flanged signs nailed to the grocer's porch post; through the era of curb signs, ornate porcelain, and even glass signs of the twenties, thirties, and forties; to plastic, internally lit modular lettering blocks showing only the brand name of products available. Time-tested trademarks were adapted, modified, or discarded. New trademarks were developed to fit the creative abilities of the plastics industry to reproduce them. The end of an era was in sight.

A second attack on the traditional gas station came in the form of no less than a presidential project. As with earlier presidents' wives, Lady Bird Johnson in the sixties selected a project to occupy her time—the result being the "Keep America Beautiful" campaign. Somehow, the roadside institution of the gas station got grouped in with unsightly billboards and roadside litter as targets in this clean-up effort. Oil companies responded with new station designs, stations built of brick that resembled the popular "rancher" style of house construction of the same era. Shell had pioneered the change in the late fifties when a city zoning board had required something "different" in station design. Though the Shell logo was still prominently featured in a rooftop pylon, the pylon was now a brick "chimney," breaking up a residential roofline. With the advent of the "Keep

America Beautiful" program, every oil company would eventually have a similar residential-looking design, proving that Pure and Phillips 66 had been on to something with their cottage designs forty years earlier.

Stations Develop an 'Image'

By the late sixties, the surviving "discounters," along with the "integrated independents" and discount operations that belonged to the major oil companies, began to innovate. Operating with smaller budgets than the majors, they didn't get caught up as much in the "residential" station look as the majors, but they experimented with image factors that would increase customer attention in other ways. Kerr McGee added canopies to many of their Deep Rock stations. Tenneco began to experiment with an early version of the convenience store. A group of innovative Spur dealers brought self-service into mainstream gasoline marketing. Gasamat in Colorado and U-filler-up in North Carolina experimented with unattended fueling using coin-operated equipment, forerunner of today's cardlocks. Crown, having purchased Washington, D. C.-area discounter Peoples, eliminated their jobber marketing and converted their entire marketing to Peoples-style super stations, operated by salaried personnel, with multiple pump islands laid perpendicular to the road under huge canopies.

Ashland began to "collect" private brands, buying out those companies that bought unbranded gas from them and had proved successful at marketing. No one knew it at the time, but each marketing move made—by the majors, by the independents, and by the discounters—would prove important in the next decade.

In the fall of 1972 Esso, Enco, and Humble dealers, some by then operating a "split island" station (one full-service island and one self-service island with multiple service bays), attached a cardboard sign to the poles supporting their identification signs. On the sign was the Esso tiger mascot holding a poster reading "we're changing our name," and an Exxon sign. The long battle of the "Standard" brand was over. When Standard had been broken up in 1911, each Standard regional marketer had been assigned a particular marketing territory. Never dreaming of competing, each used the brand name "Standard" within that territory. When the marketers expanded their territory, they were forbidden by various court rulings from using the Standard name outside their assigned areas. Each responded by adopting other trademarks. Standard of California used the Calso and Chevron brands; Standard of Indiana used the brand names Amoco, Pan-Am, and Utoco from companies they had purchased; and Standard of New York became Mobil.

Standard of New Jersey was using Enco (which replaced Carter, Pate, and Oklahoma in 1960) and Humble in areas where Esso (derived from the abbreviation "S.O." for "Standard Oil") was forbidden. A series of court rulings in the sixties, stemming from the purchase of Standard Oil of Kentucky by Standard of California and the subsequent entry into Standard (KY) markets by the Esso brand, resulted in search for a replacement brand name for all of the Standard (NJ) affiliates. After a three-year test, the most expensive rebranding in marketing history was begun as the Exxon signs were installed. Many look at this move, the last major event in the petroleum industry before the 1973-1974 "oil shortage," as the end of the classic era of service stations. Nothing would ever be quite the same.

Twenty-plus years as an oil industry observer and historian qualify me to make the following statement: The 1973-1974 oil shortage (and to a lesser extent, the 1978 shortage) was largely a hoax. The major oil companies, controlling most of the world's supply of refined product, set out to eliminate as much of the competition as possible. Independents without refineries, or with little crude production for their small refineries, were dependent on the majors for supply. And the innovative independents of the sixties were hurting many majors in marketing. The "price wars" had been expensive. So a supposed shortage gave the major oil companies an opportunity to withhold supply to many independents; with a limited supply of crude or refined products, the independents' prices would rise and eventually they would cease to operate. A convenient by-product of the shortage was a significant price increase for majors—who were paying their branded jobbers to store gasoline in their tanks while *we* were sitting in gas lines. It gave them a chance to make an extra buck, and they could even point to the independents and say, "They raised their prices first."

Many longtime service station dealers tired quickly of the hassles created by the "shortage," and got out of the business as soon as possible. Stations were closed in record numbers. Those who survived began to depend on cost-cutting measures, such as an all-self-service operation, to make a profit. Many began to add non-automotive items, such as soft drinks by the case and cigarettes by the carton, to expand their profit base. Several independents who had closed locations during the worst of the shortage began to recover, and when they did, they converted locations to what had come to be known as a convenience store.

Stations and 'C-Stores' Today

Tenneco had stores in the late sixties. Ashland had purchased the Erickson spin-off, SuperAmerica, in 1970. The convenience store was not new: Remember, the Totem stores that grew up to be Seven-Elevens had been founded in 1927. But this new breed of store was gasoline-based with store sales as an extra. Many of the earliest stores were small, but designs began to evolve. Remember the Clark regional managers, the Spur dealers, or the Texaco jobbers who in 1960 looked at Holiday stations, shook their heads, and said, "it'll never catch on"? Well, they were back for another look. More and more gasoline "C-stores" began to look much like the Holiday stations had looked years before. And the affiliated brand that Erickson's had sold off, SuperAmerica, was adding new locations and leading the way.

Today a motorist may pull up to a three- or four-grade "multi-product dispenser," get out, and notice the improved lighting from the halogen lights mounted in the overhead canopy. He may fill his car and, by-passing the pump's built-in credit card reader, enter the store for additional purchases. A menu board would let him select from deli foods or, in the case of a new trend, popular favorites from a nationally franchised fast-food program. If he opts not to place a food order, he may select from perhaps twelve brands of soft drinks dispensed from a self-service fountain, pick his headache remedy from a selection of health and beauty aids, or his cigarettes from an assortment of generics private-labeled for the store chain.

And while the canopy signage outside may reflect a nationally marketed gasoline brand, the store trademark displayed on the building or a free-standing sign may be that of a growing number of "local" jobbers or regional store operators, well-known and respected in their home community and beyond. Gasoline marketing has finally come full circle.

Acknowledgments

The authors wish to thank the following individuals and organizations for the use of their photographs in the creation of this book: Jose Ashley, Jim Bernard, Thornton Beroth, John Chance, Harold Davis, Dick Doumanian, Dave Ellis, Chip Flohe, Scott Given, Joe and Judy Gross, Lee Huff, Bob Hull, Ron Johnson, James Kelley, Bob Kaufman, Tony Lewis, Carl Mantegna, Jim Masson, Dan McGloughlin, Chris McKee, Bob Mewes, Norm Oakes, Tom Proffitt, Erwin Schmidt, Bob Scism Jr., Patrick Shimmin, Mike Thibaut, Edwin Wharton, Wendell White, Walter Wimer Jr., R.V. Witherspoon, Larry Witzel, Holiday Stations, Rex Oil Company (Stewart Kennedy), Tankar Stations Inc. (Buck Stanton), University of Alabama, and University of Louisville.

A special thanks to Debbie Henderson for all the photographic copy work. Also a special thanks to Jerry Hankins of Film Box of Greeneville, Tennessee, and Jason Wells of Wells Kingsport Camera, Kingsport, Tennessee, for all of the specialty photographic work that went into this book. We couldn't have done it without you.

Copies of many of the photographs are available from their original owners. Contact the authors for further information.

Wayne Henderson
20 Worley Road
Marshall, NC 28753

Scott Benjamin
P.O. Box 611
Elyria, OH 44036

Chapter 1

1910–1918: The Early Years— Curb Pumps and Country Stores

As Henry Ford's Model "T" began to put America on wheels, enterprising merchants of every type imaginable took notice and pondered the many ways to serve a new type of customer—the motorist. Early automobiles had many needs that frequently had to be met, including lubrication, repair, tire replacement, tire servicing, and gasoline, and it was gasoline that most frequently needed to be replenished.

Hand-operated pumping and dispensing equipment was developed along with safe underground storage, thus offering existing businesses an excellent chance to capitalize on this frequent need. Neighborhood grocers and druggists soon added curb pumps; hardware stores and rural general merchandise stores that were already selling gasoline in cans from bulk storage made the switch and the curb pump became a common sight all across America. In this chapter we will see some of the earliest examples of the most repeated commercial enterprise of this century, the service station.

Our story starts with this simple Standard station, selling gas at curbside in Canton, Ohio, in 1910. Standard had established sales offices in every town of any size to sell fuels and lubricants. Prior to the turn of the century gasoline sold throughtout these outlets was used as a specialty fuel—for cookstoves and such. The automobile brought about a gradual change of the product line and gasoline soon surpassed kerosene as the primary product. While many sold gasoline only in containers or from bulk storage, by 1910 many had added a simple curb pump, dispensing product pumped from underground or basement tanks. With the breakup of Standard in 1911, the sales offices were assigned to Standard companies in their respective areas, with the outlet in the photo going to Standard of Ohio. The Standard affiliates had to scramble around 1910-1915 to keep up with well-established chains of "filling stations" being developed by non-Standard companies in their marketing territories.
Courtesy of Joe and Judy Gross

In spite of a recent snowstorm, the underground tanks are being refilled at this Decatur, Alabama, store dispensing Texaco gasoline from a curb pump in 1918. *Scott Benjamin Collection*

Red Crown pumps were a common scene on main streets throughout Standard's marketing territory in the early twenties. Here a local tire dealer has expanded into gasoline sales and appears to be doing a brisk business. *Courtesy of Carl Mantegna*

Shackleford Auto Company, Ford dealers in Hampton, Virginia, followed the then-current trend and installed a Bowser "Red Sentry" curb pump about the time this 1918 photograph was taken. *Wayne Henderson Collection*

Dixon and Loynachan were operators of this canopied curbside Marathon station in Knoxville, Iowa, around 1916. The Marathon brand was introduced by Pittsburgh-based Riverside Oil in 1915 and was expanded greatly after being purchased by Transcontinental Oil in 1920. *Courtesy of John Chance*

Chapter 2

1918-1929

The Filling Station—Standardized Designs and The Two-Post Canopy

By the end of World War I, automobiles had proven their usefulness in everyday life and were becoming a common sight in most parts of the country. Gas station design was evolving as well, with more and more oil companies selecting standardized station designs to create a marketing "image." Although the details of each outlet were different, the new "filling stations"

Regular grade Sinclair Gasoline and a premium Benzol Motor Fuel were sold at this classic Sinclair station in 1921. *Courtesy of Erwin Schmidt*

A leaded-glass Texaco gas pump globe identifies this roadside garage and Texaco station in Hampton, Virginia, in 1922. *Wayne Henderson Collection*

Hawkeye Oil products were sold from this station in Fairmont, Minnesota, about 1922. *Courtesy of John Chance*

usually shared some common elements of design. Many stations were quite small, with narrow frontages and an integral canopy supported by two brick posts on the pump islands. Gas pumps positioned either inboard or outboard of the canopy posts served to create a "showcase" effect, particularly when lit up at night.

Also in this decade the automobile dealership or repair garage became a more important outlet, often becoming the only place in town where curb pumps were "accepted" in terms of traffic flow. Brand marketing became more important in this era, and station signage reflects this new importance. Enterprising independent dealers were still free to create eye-catching fantasy stations, with various unlikely subjects (lighthouses, windmills, and such) being chosen for station design. Also of note are the many "cottage" designs, often more commonly associated with stations of the thirties, that made their first appearances in the late twenties.

This Hawkeye bulk plant and filling station was serving Mason City, Iowa, about 1922. *Courtesy of John Chance*

Hawkeye Oil was operating the Red Ball Filling Station in Waverly, Iowa, about 1922. *Courtesy of John Chance*

This painted fence served as an elaborately painted billboard advertised Hawkeye "Red Ball" gasoline and "Faultless" oils. *Courtesy of John Chance*

This neighborhood station was selling Hawkeye products in Clear Lake, Iowa, around 1922. *Courtesy of John Chance*

This retouched promotional photo was used to advertise the Hawkeye "Red Ball" stations. *Courtesy of John Chance*

Another automobile dealer has added gasoline to his product mix. Shown here is the Chevrolet and Hudson/Essex dealer in Belle Plaine, Iowa, selling Hawkeye gasoline around 1922. *Courtesy of John Chance*

A curb sign identifies this Phoebus, Virginia, station as a Sinclair station in 1922. *Wayne Henderson Collection*

K-T Oil Company dealers gather for a group photo in front of the Oakley, Kansas, K-T station in November, 1923. Later the K-T brand would evolve into Elreco. *Courtesy of Jim Masson*

Standard "Red Crown" products were sold from this curb pump at the Overland/Willys-Knight dealer in Berne, Indiana, in 1923. *Courtesy of Carl Mantegna*

Cosby's Garage in Hampton, Virginia, was selling Texaco gasoline from a curb pump, along with new Chevrolets and automotive repair when photographed here in 1923. *Wayne Henderson Collection*

This Peerless station in Dayton, Ohio, in 1923 is one of the early type of stations that were of prefab construction. Station kits could be ordered from steel building manufacturers and assembled on site in a day or two. *Courtesy of Bob Hull*

This auto repair garage on Mellen Street in Phoebus, Virginia, was selling Texaco products from curb pumps in 1924. *Wayne Henderson Collection*

Welcome to Mayberry, USA. For many years this station offered Standard (Esso) products at the south edge of Mt. Airy, North Carolina, on US601. The fictional Mayberry, North Carolina, was patterned after Mt. Airy, and perhaps the station shown in this 1925 photo was the inspiration for Mayberry's "Wally's Garage." *Courtesy of Thornton Beroth*

Driftwood, Oklahoma, was home to this tiny roadside filling station around 1925. *Courtesy of Carl Mantegna*

Harbor City, California, was home to the Pine View Service Station. Shown here is a progression of brands, from a dual-branded Pan-Gas/Gilmore, to the addition of the Richfield brand, and finally offering Richfield, Union, Gilmore, and El Camino products. Multiple brands were a feature of many California stations. *Courtesy of Carl Mantegna*

White Rose Gasoline was delivered to this station in a brand new 1925 Federal tank truck. The location is unknown, but it is probably in Ohio. *Courtesy of Carl Mantegna.*

These three photos, taken on the corner of Cranston and Dodge streets in Providence, Rhode Island, show the transformation from local "jobber" brand to major brand. In the first scene, from about 1925, the small station building has a sign that reads "Lamson Oil Company" and pump globes advertise Lamson's "Nun-Bet-Er" brand. The second scene, from 1929, shows the conversion to Richfield of New York in progress; note the Richfield signage with the Nun-Bet-Er globes. The third scene, from March 1936, shows a larger station building, the brand new Richfield of New York shield sign, and a complete elimination of the Lamson name. *Courtesy of Thornton Beroth*

Providence, Rhode Island, Richfield jobber Lamson Oil Company was operating this station under their own "Nun-Bet-Er" brand when photographed here around 1926. *Courtesy of Thornton Beroth*

This unusual mushroom-shaped station was being operated by Gay Oil Company in Little Rock, Arkansas, when photographed here about 1926. *Courtesy of Carl Mantegna.*

"How About Gas?" reads the one-piece etched globe atop the unusual twin visible pump at this Lamson Oil Company station in Providence, Rhode Island, in 1926. *Courtesy of Thornton Beroth*

A fantastic "Rotary" lift made service work easy at this Associated station in California in 1926. A car could be easily rotated for access to points needing lubrication or repair. *Courtesy of Carl Mantegna.*

Hawkeye Oil marketed outside Iowa and Minnesota under the "Parco" brand. This Hawkeye Oil "Parco" station was located in Greensburg, Indiana, about 1926. *Courtesy of John Chance.*

Hawkeye Oil joined IOMA in 1926, as is in evidence from the Independent Oils eagle logo in this 1926 Paris, Kentucky, photo. *Courtesy of John Chance.*

Lighthouses were the perfect attention-getter for service station structures located hundreds of miles from navigable water. Medicine Lodge, Kansas, was home to this beacon-turned-Kanotex station in the late twenties. *Courtesy of Jim Masson.*

This Standard (Indiana) "Red Crown" station in Robinson, Kansas, features only a single pump, but had ventured into automotive accessories as the local Kelly tire dealer. Although most major brand stations did not offer TBA (tires, batteries, accessories) prior to 1930, the success of tire stores venturing into gasoline sales, as well as the elaborate TBA marketing by several independents and most local jobbers, led the way for the majors to get involved in TBA and auto repair. *Courtesy of Patrick Shimmin*

Gulf products were sold at this unusual station at The Bottle, Alabama, (outside Opelika) from 1924-1935. The station, created as a promotion for Nehi (based in nearby Columbus, Georgia), attracted the attention of everyone who happened to pass by. The station, which was rebuilt much more modestly after a 1935 fire, is shown here in about 1928. *Courtesy of James Kelley and University of Alabama*

Hop's Place, Virginia's oldest auto parts and repair business, was selling Gulf gasoline when photographed here in 1928. *Wayne Henderson Collection*

Rooftop Sinclair globes and "tall ten" pumps are features of this classic service station on US 20 in Conneaut, Ohio, in 1928. *Courtesy of Ed Wharton*

This Hampton, Virginia, Amoco station was identified by an ornate "Amoco Gas" internally lit glass sign when photographed in 1928. Note the unusual air stand. *Wayne Henderson Collection*

Pennsylvania motorists were offered three grades of Atlantic Gasolines—Atlantic, Atlantic Ethyl, and Atlantic 68-70—from curb pumps in this 1928 photo. *Courtesy of R.V. Witherspoon*

Cities Service Regular and Koolmotor Premium gasolines were offered from this tiny Cities Service station in Conneaut, Ohio, about 1928. *Courtesy of Edwin Wharton*

Walburn Oil Company of New York City was merged with California's Richfield Oil to form Richfield of New York. This Lamson Oil Company station on Olneyville Square in Providence, Rhode Island, was converting to the Richfield brand in 1929 when this photo was taken. Note the Walburn Ethyl globe on the Richfield Ethyl visible pump in this scene. *Courtesy of Thornton Beroth*

Beginning in 1929, Standard of Ohio built a series of stations in English Tudor design. These stations, in high visibility marketing areas, attracted a great deal of attention from the competition and likely inspired a trend in improved station design. This location is thought to be in Cleveland and was photographed in 1929. *Courtesy of Joe and Judy Gross*

Curb pumps at a Main Street garage in Grafton, Ohio, dispense Standard (Ohio) Red Crown regular and Sohio Ethyl premium gasolines in 1929. *Courtesy of Joe and Judy Gross*

Twin pump islands with a central station under the same canopy. Not exactly a new design, as is shown by this Union Oil station in California photographed in 1930. *Courtesy of Carl Mantegna*

The Sohio brand name was introduced in 1928 to consolidate Standard of Ohio's marketing under a single brand. This "Type A" station is an example of an early Sohio standardized station design and sports early Sohio branding from about 1930. *Courtesy of Joe and Judy Gross*

The State Line tourist camp at East Conneaut, Ohio, was dispensing Sinclair products when photographed in this postcard view in 1930. *Courtesy of Ed Wharton*

Tourist cabins were a logical evolvement for the roadside service station operator. This Shell dealer was located along Rt. 20 in Conneaut, Ohio. *Courtesy of Ed Wharton*

This elaborate Sunoco station in Alliance, Ohio, was dispensing Sunoco gasoline, Atlantic gasoline, and Mobil oil products when photographed here in 1929. *Shawn Watson Collection*

Chapter 3

1929-1942

The Depression—Gas Stations Come Of Age

The Depression era gave hundreds of people a most unusual opportunity. While most people would not have given up a steady job to chase a dream of owning their own business, those who were put out of work by the Depression economy often had nothing left to lose and gambled the several hundred dollars necessary to start their own business—a service station. Fortunately, many succeeded, at least for a time, and in this chapter we see the evidence of their success. Odd

Free air came from a shell-topped curbside air stand at this Shell station photographed about 1930. *Courtesy of Carl Mantegna*

Richfield's Ethyl grade was being introduced when this station on Elmwood Avenue in Providence, Rhode Island, was photographed in 1930.
Courtesy of Thornton Beroth

Union Oil products and Bell Gasoline (no relation to Bell Oil of Tulsa) were sold by this station in British Columbia about 1930. *Scott Benjamin Collection*

Above and right
Main Street in Marshall, North Carolina, for many years along the main truck route from all points in North Carolina to Knoxville and beyond, was home to the French Broad Tea Room and Service Station, offering Standard Regular and Esso Ethyl when photographed here in 1930. *Wayne Henderson Collection*

lots in big-city industrial areas, small town street corners, highway locations between somewhere and somewhere else, surrounding every kind of tourist attraction imaginable—gas stations were suddenly everywhere.

Most major oil companies eliminated direct, salaried operations in the early thirties, both to save operating expenses and to avoid chain store taxes. Anxious former station employees and many newcomers to the trade became "dealers" for national and regional brands. Trackside discounters, an invention of the twenties, gained a much larger share of the gasoline market in the cost-conscious thirties and major oil companies introduced "third" grade products, low-octane gasolines with no tetraethyl lead meeting "U.S. Motor" specifications, in order to compete with the tracksiders.

Brand images were improved and station design became more of an industrial "package" in this era. Automotive repair, beyond simple lubrication, became commonplace at the neighborhood service station. The photos chosen for this section will show that many of the nostalgic images we have of the gas station come from this era.

At this Marshall, North Carolina, Shell station in 1930, Pennsylvania grade oils under the Pennzoil and Wm Penn brands were offered alongside Shell products. The young attendant beside the pump later became the dealer at this station and operated a station on this site until his death in the nineties. *Wayne Henderson Collection*

Shell products were sold at this tiny station in Buyne Fall, Michigan, in 1931. *Courtesy of Tom Proffitt*

Neon Richfield signage attracted attention to Mack's Community Service Station, somewhere on the West Coast about 1932. *Courtesy of Carl Mantegna*

These young ladies are filling up their Packard with Kendall gasoline in Pennsylvania in 1932. *Courtesy of Carl Mantegna*

"Follow The Golden Trail" to Richfield products on Cranston Avenue in Providence, Rhode Island, about 1932. *Courtesy of Thornton Beroth*

This former Colonial Beacon station had rebranded to Richfield by the time this 1932 photo was taken. *Courtesy of Thornton Beroth*

Depression-era economy grade Richfield "Blue Streak" was a featured product at this New England Richfield station about 1932. *Courtesy of Thornton Beroth*

Union "76" was a new Union Oil product marketed by this unique Union Oil Company station in British Columbia in 1932. *Scott Benjamin Collection*

This Spur station, on South Third Street in Louisville in 1933, is an example of an unusual trackside station. Although the station was built on the corner of the existing business block, tanks were filled from a rail spur line that ran nearby. The elaborate painted backdrop made the entire station into a huge billboard advertising Spur products. *Courtesy of University of Louisville*

"Partners in Power—Richfield and Richlube" is the oft-repeated theme at this classic Richfield station on Columbus Avenue in Springfield, Massachusetts, in 1933. *Courtesy of Thornton Beroth*

Notice the unusual "Sinclair Oils" oil can rack, complete with globe, at this 1934 Sinclair station in Hampton, Virginia. *Wayne Henderson Collection*

White Eagle globes on the roof and a cast iron White Eagle at curbside identify this combination Ford dealership and White Eagle station in Mott, North Dakota, in 1934. *Courtesy of Bob Mewes*

Traffic congestion had all but eliminated the curb pump by the time this garage selling Pure "Purol Pep" was photographed in the mid-thirties. *Courtesy of Carl Mantegna*

British American "B-A" products are sold from an unusual twin visible pump at curbside by this variety store in Quebec about 1934. *Scott Benjamin Collection*

This close-up shows how gas pumps were built into canopy columns in the Sohio English Tudor stations. These are not at all unlike today's multi-product dispensers built to blend into canopy columns. Note the globe-topped oil cabinet. *Courtesy of Joe and Judy Gross*

Sohio products were sold through hundreds of stations like this throughout Ohio in the thirties. *Courtesy of Joe and Judy Gross*

Associated "Flying A" products are offered at the elaborate Miller Service Co. Firestone tire store in California about 1934. *Courtesy of Dave Ellis*

Multiple brand stations were a California phenomena, taking advantage of laws in that state that allowed dealers to be truly independent and represent numerous suppliers. This station was selling Sunset, Union, and Gilmore products when photographed in 1934. *Courtesy of Dave Ellis*

This prefabricated structure became a standardized style of Union station in the early thirties and hundreds were built until World War II. *Courtesy of Dave Ellis*

Gilmore products were often sold through stations featuring elaborate signage. This particular station, though, is decorated in a tastefully functional way. *Courtesy of Dave Ellis*

Shell was also represented in California through a series of elaborate prefabricated stations. This location was photographed in the mid-thirties. *Courtesy of Dave Ellis*

Streamline design was another feature of California stations. This Union Oil location featured sweeping curves to either side of a central office. *Courtesy of Dave Ellis*

The open-air look of California stations was accomplished by the use of carport service facilities and projecting canopies, often with rooftop signage. This Union station from the late thirties is an excellent example. *Courtesy of Dave Ellis*

Even Texaco, in the days before the Teague station design was adopted, used examples of open air design. These two stations were formerly stations of the California Petroleum Co. before converting to Texaco in the mid-thirties. *Courtesy of Dave Ellis*

Rooftop pylon signage attracts attention to this typical Gilmore station in 1934. *Courtesy of Dave Ellis*

A rooftop lion and an unusual angled service area are featured at this Gilmore outlet in California in 1934. *Courtesy of Dave Ellis*

This classic Richfield station was located in southern California in 1934. *Courtesy of Dave Ellis*

The Richfield of California image evolved into something distinctly different from that shared with Richfield of New York as Richfield (California) recovered from bankruptcy in the late thirties. *Courtesy of Dave Ellis*

This small Richfield outlet was offering economy grade "Fleet" among its products in 1934. *Courtesy of Dave Ellis*

Rio Grande was a Sinclair affiliate operating on the West Coast. When Sinclair purchased a substantial share of Richfield (California), including all of Richfield of New York, during their bankruptcy, Rio Grande was assigned to the revitalized Richfield as a secondary brand, surviving into the seventies. *Courtesy of Dave Ellis*

Associated introduced the famous "Flying A" brand as an aviation-quality regular grade in 1932. After their merger with Tydol, Flying A became their primary brand. This location was dispensing Associated products in 1934. *Courtesy of Dave Ellis*

Seaside was a long-time Associated secondary brand, marketing throughout California. This location was photographed in the mid-thirties. *Courtesy of Dave Ellis*

This tiny office structure (with the garage behind) was a direct predecessor to the seventies' island kiosk at self-serve pumpers. *Courtesy of Dave Ellis*

Notice the Red Lion ad glass panel in the regular grade "Roar With Gilmore" Bowser pump at this California Gilmore station in the mid-thirties. *Courtesy of Dave Ellis*

California-based Hancock sold gasoline through a series of stations like the one shown in the mid-thirties. *Courtesy of Dave Ellis*

This elaborate "Y" shaped Richfield station featured two pump islands and an open-air service area when photographed here in the late thirties. *Courtesy of Dave Ellis*

Simple but functional was this tiny Associated station in California in 1934. *Courtesy of Dave Ellis*

This elaborate multi-pump location, perhaps the inspiration to numerous postwar multi-pumps, was dispensing several brands of gasoline when photographed here in 1934. It was located at Second and LaBrea in Los Angeles. *Courtesy of Ron Johnson*

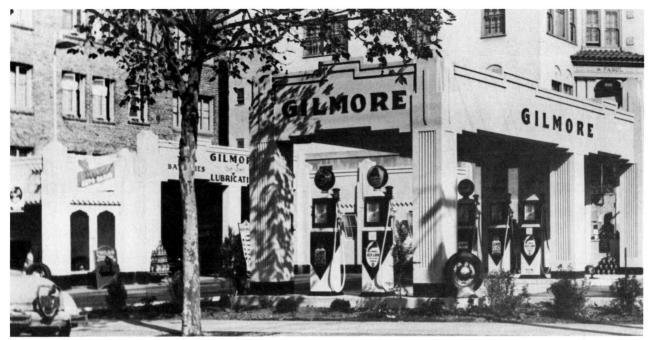

A curbside tree shades this beautiful Gilmore location in Oakland, California, in the late thirties, a full twenty years before another California city required the use of trees and shrubbery as well as eye-pleasing design at a Shell outlet that heralded the introduction of the rancher-style stations. *Courtesy of Ron Johnson*

The 14 1/2 cents per gallon wouldn't even pay the taxes on a gallon of Super Shell nowadays, but was an attractive price to the 1934 motorist passing this Shell station in Conneaut, Ohio. *Courtesy of Edwin Wharton*

57

The Dodge and Plymouth dealer in Marshall, North Carolina, in 1935, like most of the automobile dealerships in the country, was selling gasoline in addition to new and used cars. All that remains today of this old-time car dealer is the massive signpole supporting the Esso sign in this photograph. *Wayne Henderson Collection*

An Atlantic Station in Miami, Florida? Well maybe not. How about: An Atlantic station on Rt. 6 in Linesville, Pennsylvania, in the winter of 1935. *Courtesy of Edwin Wharton*

This series of photos shows station progression over a period of fifteen years. The first, taken about 1935, shows a station of non-standardized design, complete with thirties-era Gulf "Authorized Dealer" signs and a visible pump. The second scene, from September 1937, shows the station reimaged with computing pumps and a Gulf "Type K" double mast-arm sign pole. The later scene is the same location, obviously having prospered enough to have been rebuilt in a Gulf standardized design. The station was located at 2600 Magnolia Ave. in Knoxville, Tennessee. *Courtesy of Chip Flohe*

It was a rough winter for Atlantic stations. These curb pumps were dispensing Atlantic products (or at least attempting to), to passers by (few in number) in Albion, Pennsylvania, about 1935. *Courtesy of Edwin Wharton*

Six pumps were dispensing Richfield Gasoline at this Pawtucket, Rhode Island, cottage-type station in 1935. *Courtesy of Thornton Beroth*

Time for spring cleaning at this Indianapolis Criteria station in 1935. Criteria was an affiliate of the Indianapolis based discounter Wake Up. *Courtesy of Jim Bernard*

This Indianapolis station is one of the first Wake Up locations. Wake Up was an Indianapolis-based independent, operating stations throughout central and southern Indiana. *Courtesy of Jim Bernard*

A motorist fills up with Criteria gasoline in Indianapolis in 1936. *Courtesy of Jim Bernard*

Trackside and discount operators often built stations on odd-shaped city lots like this tiny Criteria station in Indianapolis about 1936. *Courtesy of Jim Bernard*

This is the classic Criteria station where the earlier photos of spring cleaning and a motorist filling up were taken. Shown here about 1936 in Indianapolis. *Courtesy of Jim Bernard*

Pennzip gasoline and Pennzoil motor oils are the featured products at this tiny station in Pennsylvania in 1936. *Courtesy of Bob Kaufman*

Tokheim 36B pumps feature illuminated tops advertising Richfield products at this station at 645 N. Fourth Ave. in Los Angeles, California, in September, 1936. *Courtesy of Ron Johnson*

This historic Skelly station stands alongside Union Station in Kansas City, Missouri, (at 220 W. Pershing Road) and is built in classic federal style typical of the surrounding downtown buildings. A neon-illuminated spire with Skelly trademarks and thermometer serves as an eye-catching art deco contrast to the rest of the building in this 1936 photo. The spire was later replaced by a much larger rooftop neon Skelly diamond sign. *Courtesy of Jim Masson.*

This historic station on South Valley Street in Kansas City, Kansas, was originally a residence built of native stone and wood. It has seen a progression of brand changes over the years. In the early photo, dating from about 1936, Kanotex products are sold alongside those of Kansas City jobber Winstun Oil. A Winstun sign identifies the station. In the next scene, probably 1940, a neon Skelly sign announces the brand changeover to Skelly products. In the final scene, from 1953, Sinclair products are offered at Winters Service. *Courtesy of Jim Masson.*

Standard of California marketed direct to the public through the chain-operated "Standard Stations, Inc." locations, as well as through dealers labeled as "Authorized Distributors." This dealer location was photographed about 1936. *Courtesy of Dave Ellis*

Ornate signage and a dignified look were typical of gas station design in the twenties as is shown in this photo of a classic sandbrick Gulf station. This station was still operating in Clarksville, Tennessee, in 1936. *Courtesy of Chip Flohe*

The Belle Meade Gulf Service is another example of streamline service station design. The extensive use of glass panels gave the station an open-air design during the day and transformed the station into a display showcase, complete with neon facade lettering, at night. The station was constructed about 1936 in the Nashville suburb of Belle Meade and is shown here in a postwar photo. *Courtesy of Chip Flohe*

The classic "sand brick" Gulf was named for the sand color brick in the canopy columns that accents the dark brown brick used to build the rest of the station. This was the first standardized gas station design, introduced in 1916. The last ones were built about 1932 and many still remain in use today. This one was located in Tennessee in 1937. *Courtesy of Chip Flohe*

This roadhouse and grocery store was selling Mobil's economy grade "Metro" gasoline in 1937. *Courtesy of Mike Thibaut*

Virtually the entire California petroleum industry was represented at Herald Pilson's station on West Florence in Downey, California, when it was photographed in November 1937. *Courtesy of Dan McGloughlin*

By 1937 Hop's Place had rebranded Amoco and the service station in the central courtyard of the U-shaped building had been modified into a canopy. A year later the station had been reconstructed (there's a photo of it later in this chapter) and would offer Amoco products to Hampton, Virginia, motorists until 1970. *Wayne Henderson Collection*

Wilshire (Polly/Economy) and Reliance products were offered at A.L. Pautz's station at Jefferson and Western in Los Angeles when photographed here in September 1937. *Courtesy of Dan McGloughlin*

This elaborate Shell station, typical of many California super stations, was located at 5857 Sunset Boulevard in Hollywood when photographed in January 1937. *Courtesy of Ron Johnson*

Gilmore's racing ties were suggested by the flagman atop this Long Beach, California, Gilmore station in August 1937. *Courtesy of Ron Johnson*

Rooftop signage at this Mohawk station features illuminated globe lenses in frames, plugged into a ceiling fixture. Mohawk gasoline was being dispensed from a twin clock-face pump when this 1937 photo was taken. *Courtesy of Dan McGloughlin*

York Oil Company, the Conoco jobbers in Hampton, Virginia, was operating this classic station on East Queen Street in Hampton in 1938. *Wayne Henderson Collection*

Linco had began the gradual conversion to Marathon when this photograph was taken in Eastern Illinois around 1938. Marathon motor oils are offered, as well as Marathon's economy grade "Multipower" gasoline. *Courtesy of Edwin Wharton*

Named for the city where it was developed, this classic Amoco station is an example of what is known as the "Baltimore style." This station was photographed in 1938 at Wythe in Hampton, Virginia. *Wayne Henderson Collection*

Notice the Wayne Model 492 "Roman Column" pumps at this Murray, Kentucky, Gulf bulk plant in 1938. *Courtesy of Chip Flohe*

This Amoco station was equipped with twin outside grease racks and identified with numerous painted billboards advertising Amoco products when photographed in 1938. It still operates near Hilton Village in Newport News, Virginia. *Wayne Henderson Collection*

Another example of the classic Amoco "Baltimore" style station is this location photographed in 1938 at 28th Street and West Avenue in Newport News, Virginia. *Wayne Henderson Collection*

This roadside restaurant was selling Amoco products to motorists passing between Warwick Courthouse and Yorktown, Virginia, in 1938. *Wayne Henderson Collection*

Numerous bulk plants, such as the Amoco plant shown here, along the "Small Boat Harbor" in Newport News, Virginia, were offering marine fueling services in 1938. *Wayne Henderson Collection*

Phoebus Motor Company, Plymouth and DeSoto dealers in Phoebus, Virginia, was offering Amoco products when photographed in 1938. *Wayne Henderson Collection*

Hop's Place, a Hampton, Virginia, landmark garage founded in 1922 and this writer's former employer, had rebranded from Gulf (see the photo in chapter 2) to Amoco by the time this photo was taken on South Armistead Avenue in 1938. *Wayne Henderson Collection*

Texaco Sky Chief gasoline was a brand new product when this Texaco station in Memphis—complete with Wayne Model 60 pumps offering both Fire Chief and Sky Chief—was photographed in 1939. *Courtesy of Chip Flohe*

Built in the twenties and branded White Rose at the time of this 1939 photograph, this station on Mississippi Boulevard in Memphis had rebranded Mobilgas by the time this 1955 photo was taken. *Courtesy of Chip Flohe*

The Thomas Street business district was home for this classic Gulf station in Memphis in 1939. *Courtesy of Chip Flohe*

It can only be assumed that the jobber for White Rose products in Memphis had rebranded Mobilgas by the early fifties as these two photos show a classic 1939 White Rose Station, complete with Boy and Slate at the curb, and the same station about 1953, branded Mobilgas and identified with a rare neon Mobilgas shield sign. This station was located on Chelsea Avenue at Breedlove in Memphis. *Courtesy of Chip Flohe*

The Thomas Street Garage, sharing a middle-of-the-block location with a Kroger store and a farm supply store, was offering Pan-Am products when this 1939 photograph was taken in Memphis. *Courtesy of Chip Flohe*

This 1939 Memphis Gulf station featured an outside wash rack and ornate Gulf sign structure that was used only at their highest volume stations. The 1955 photo shows the same station not nearly so ornate. Changing traffic patterns and outdated design had obviously taken their toll on this location. *Courtesy of Chip Flohe*

This Esso station was equipped with twin pump islands and attached canopies when photographed at Seventh Avenue and Chelsea Avenue in Memphis in September 1939. By the time the later photo was taken in 1954, the ornate twenties station was gone and a typical porcelain box station had taken its place. *Courtesy of Chip Flohe*

Gannaway and Diggs were busy Shell dealers in Hampton, Virginia, in 1939. The business is now known as Dixie Diggs Auto Parts and has not sold gasoline in over thirty years. *Wayne Henderson Collection*

This 1940 Gulf station is an excellent example of one of the earlier Gulf "icebox" station designs. *Wayne Henderson Collection*

Spur was still marketing only one grade of gasoline when this 1940 photograph was taken in Louisville. The spacious driveway, well-stocked oil racks, and tasteful signage shows a gradual refinement of the discount gasoline image. *Courtesy of University of Louisville*

The automobile in this photo was over twenty-five years old by the time this Texaco station was photographed about 1940. *Courtesy of Scott Given*

One of the earliest Leonard branded stations was this 1940 Kalamazoo, Michigan, location. The early Leonard "flag" logo appears here in neon. *Courtesy of Chris McKee*

C.F. Tuttle was a Kingsville, Ohio, Amoco jobber, operating out of this office and filling staiton about 1940. *Courtesy of Edwin Wharton*

Tuttle supplied this station in Jefferson, Ohio, with Amoco products. *Courtesy of Edwin Wharton*

This classic Amoco station, at US 20 and Mill Street in Conneaut, Ohio, still displayed the twenties Amoco logo when photographed about 1940. *Courtesy of Edwin Wharton*

Seen here is a location in Ashtabula, Ohio, where Amoco jobber C.F. Tuttle converted residential locations into Amoco stations. *Courtesy of Edwin Wharton*

Birmingham-based Mutual Oil was operating this location in Sheffield, Alabama, in 1941. Note the "free premiums" display and the Crown "Greenzoil" motor oil cans. *Wayne Henderson Collection*

This "jobber brand" Pow-R station was photographed alongside US 52 just north of Winston-Salem, North Carolina, in 1941. *Courtesy of Lee Huff*

This modified main street storefront service station was selling Pure Oil products in Mars Hill, North Carolina, in 1941. *Courtesy of Lee Huff*

Country stores grew up around rural crossroads beginning about the end of the Civil War. Many of these crossroads stores added gasoline and oils to their varied product mix in the early years of this century. This store outside Reidsville, North Carolina, appears to be a vintage-1900 location and was selling Pure Oil products when photographed here in 1941. *Courtesy of Lee Huff*

This Farmington, North Carolina, country store, selling Pure Oil gasoline in 1941, featured a pump island canopy with rooftop advertising signs, not at all unlike the canopy at a modern convenience store. *Courtesy of Lee Huff*

Can't you just hear Roy Acuff's "Wreck on the Highway" playing on the jukebox at this Hickory, North Carolina, roadhouse and barbecue restaurant selling Pure Oil gasoline in 1941. *Courtesy of Lee Huff*

Pure used National A-38 gas pumps almost exclusively during the forties, but these are illuminated by pump-top "station lighter" attachments at this tiny Pure Oil cottage on the highway between Asheville and Marshall, North Carolina, in 1941.
Courtesy of Lee Huff

Atlanta-based Moore Oil was selling Greyhound gasoline at this station on North Church Street in Spartanburg, South Carolina, in 1941. After World War II Greyhound became Citizens 77, later a part of Bay/Tenneco.
Courtesy of Lee Huff

Registered rest rooms? Not at this Red Springs, North Carolina, country store rebranding from Richfield to Pure in 1941. The outhouse door is standing wide open, perhaps for a cleaning to qualify it for some oil company's forties clean rest room program.
Courtesy of Lee Huff

Notice the tire stack, complete with mannequin, at the elaborate Muller Bros Associated station in Hollywood, California, in 1941. *Courtesy of Ron Johnson*

Elaborate signage identifies this Montana Grizzly station in 1940. Note the three-dimensional bear logo. *Courtesy of Joe Ashley*

Chapter 4

1942-1945

The War Years—Rationing and Restrictions

Gasoline and tires were not the only things in short supply from wartime service stations. Photographs were scarce, too. In this short chapter we show several examples of wartime gas stations. The wartime station operator had to deal with uncertain supply, gasoline and tire rationing cutting into "free-market" sales, and a labor shortage that often saw young ladies taking the place of young men pumping gasoline and cleaning windshields.

Stations closed by the thousands because they became unprofitable. Thousands more closed when Depression-era station operators, forced into their marginal businesses out of necessity, suddenly took the opportunity of available wartime work in defense plants or answered Uncle Sam's call.

Shown here are two scenes of a small Illinois Marathon station. The first, from about 1940, shows the operator proudly standing in front of a brand new Wayne Model 70 gas pump dispensing Marathon regular gasoline. Marathon Ethyl is dispensed alongside from a trusty visible pump. The later scene, from about 1943, shows all electric pumps, a most unusual change for wartime, and sculptured shrubbery reading "Victory USA" across the front of the station. Many gas stations were collection depots for wartime recyclables and joined in the effort with great patriotism. *Courtesy of Harold Davis*

Without a doubt, the most familiar sight on Ohio roads in its day. Sohio had sold Esso motor oils for many years, but here we see this premium motor oil product being sold from lithographed glass bottles proudly displayed on the pump island. Glass packaging replaced steel for many products during World War II due to shortages of critical materials. *Courtesy of Joe and Judy Gross.*

Crossroads, USA. This so-typical scene was photographed in Rock Creek, Ohio, in September 1945. Gulf products are offered at curbside by a local garage operator. At war's end, travel restrictions were lifted and the corner gas station quickly came back to life. *Courtesy of Edwin Wharton*

Chapter 5

1945-1954

Early Postwar Years

Money flowed freely, as did gasoline at the end of World War II. Returning G.I.s, weary defense plant workers, and those whose travel was restricted by wartime gasoline and tire rationing, took to the highways in record numbers. America renewed its love of the open road and gas station operators everywhere collected the tolls.

The corner service station had became an automotive repair center in the Depression years, and in the postwar years dealers would learn to "merchandise" their services.

In this chapter we see gas stations greatly revitalized in the early postwar years, but there was little true change in the designs of stations in use. Oil companies knew they could sell every drop of gasoline produced, and there was little reason to worry about images.

Wooster, Ohio-based Red Head Oil Company was operating this location on Bluefield Avenue in Bluefield, West Virginia, in 1946. Red Head products were sold here for over fifty years. *Wayne Henderson Collection*

While the location of most of the sites shown in photographs in this book have been identified, the location of this scene is unknown. It can only be said that this photo represents a welcome stop for weary motorists as they approach virtually every town throughout the original twenty-nine-state Gulf marketing territory. *Courtesy of Chip Flohe*

Notice the unusual Pure Ethyl pump signage at this classic Pure cottage station in Crossville, Tennessee, in 1946. *Courtesy of Chip Flohe*

The Langley Service Station has operated on the corner of West Queen Street and Armistead Avenue in Hampton, Virginia, since 1919. The dealer who took over the station about the time this 1946 photograph was taken was dedicated to 24-hour service and for over thirty-five years the station was never closed, even during a reconstruction in 1970. *Wayne Henderson Collection*

Competitors in Crossville, Tennessee, in 1946 were these two stations—a Sinclair service station and a corner grocery store selling Texaco products. Of note is the fact that the Texaco is still displaying the small signs that had been replaced by the banjo sign in 1936. *Courtesy of Chip Flohe*

In the days before transports and vapor recovery, Esso station tanks were refilled by trucks like the one shown in this 1946 Crossville, Tennessee, scene. *Courtesy of Chip Flohe*

Sinclair products were offered by Hensley's Garage on Chapman Highway in Knoxville in 1946. *Courtesy of Chip Flohe*

In river towns a small number of bridges typically funnel traffic between sections of a city, making station locations at the ends of these bridges very desirable to take advantage of traffic in "captive" flow. This Pure cottage at the end of the Henley Street Bridge in Knoxville appears to be quite busy in this 1946 photo. *Courtesy of Chip Flohe*

Gulf stations such as the one seen here were very much a part of every main street in the southeast when this photo was taken in Crossville, Tennessee, in 1946. *Courtesy of Chip Flohe*

Trackside discounters frequently located near major brand stations to further enhance their price-competitive nature. This classic Gulf station, captured in 1946 in an odd camera angle, competes with the Martin Oil Company discount station next door seen in the background of the photo. *Courtesy of Chip Flohe*

This classic Esso station, a variation of the design Standard (New Jersey), was built around 1920, and was still in use in Knoxville in 1946. *Courtesy of Chip Flohe*

Neighborhood and roadside grocers learned the benefits of selling gasoline in the early years of this century. This roadside operation was selling Texaco gasoline in Crossville, Tennessee, in 1946. *Courtesy of Chip Flohe*

This Edgerton, Wisconsin, cafe and store was selling Sinclair products when photographed about 1946. *Courtesy of Larry Witzel*

Crown-topped pumps meant "Standard" from the teens until 1961. Shown here is a typical Standard station in conjunction with a local automobile dealer, about 1947. *Courtesy of Patrick Shimmin*

Free-standing gas station canopies are nothing new, as is witnessed by the elaborate canopy at this 1947 Pure station at the forks of the road in Columbia, Tennessee. *Courtesy of Chip Flohe*

This rare photograph shows a classic "Teague" style Texaco station in Memphis just prior to its opening day in 1948. *Courtesy of Chip Flohe*

This Raleigh, North Carolina, taxicab, advertising an Esso-sponsored news program, was filling up with Richfield gasoline in 1948. *Wayne Henderson Collection*

This roadside Shell station catered to motorists in Kingsport, Tennessee, in 1949. *Courtesy of Chip Flohe*

Tankar Stations attract customers with an extensive "Free Premiums" program, redeeming savings coupons for housewares and other items displayed in the small building adjacent to the main station building. This station was located on High Street in Portsmouth, Virginia, in 1950. *Courtesy Tankar Stations/Wayne Henderson Collection*

Wartime boom town Oak Ridge, Tennessee, was the setting for this roadside Sinclair station, shown in this 1950 photo. *Courtesy of Chip Flohe*

Complete automotive repair was available at this corner Cities Service station in Erwin, Tennessee, in 1950. *Courtesy of Chip Flohe*

A neon Shell sign and shell-shaped pump globes attract attention to this classic Shell station in downtown Erwin, Tennessee, in 1950. *Courtesty of Chip Flohe*

This Knoxville Texaco station has intact all of the visual effects that make this station design so attractive. *Courtesy of Chip Flohe*

While Pan-Am stations used the same type of sign structure as the familiar Standard signs of the Midwest, variations did exist, such as the smaller version used by this Knoxville Pan-Am station in 1950. *Courtesy of Chip Flohe*

A neon rooftop Esso sign attracts attention from across the Tennessee River to this Esso station at the end of the Gay Street Bridge in Knoxville in 1950. *Courtesy of Chip Flohe*

The dealer's name held a place of prominence in Esso station design from the thirties until the corporate name "Humble" was added to station facades in 1961. Dealer J.T. Knittel operated this Knoxville station in 1950. *Courtesy of Chip Flohe*

A streamlined, round-corner design was chosen for this Knoxville Esso station, which is shown here in a 1950 photo. *Courtesy of Chip Flohe*

This Knoxville automobile dealership sold Firestone tires and Texaco gasoline in 1950. *Courtesy of Chip Flohe*

These two photos show the upper and lower levels of a most unusual Gulf station on Chapman Highway in Knoxville in 1950. While the upper level was accessed from Maryville Pike, the lower level faced Chapman Highway. Station attendants had to run up and down the stairs all day while serving the motoring public. Note the thirties-style Gulf sign at the rear of the station. *Courtesy of Chip Flohe*

The classic design of a thirties Gulf station is still intact at this location in downtown Knoxville in 1950. *Courtesy of Chip Flohe*

Lion Oil Company operated this very dignified-looking gas station in Memphis in 1950. *Courtesy of Chip Flohe*

It seems that no two Pure cottage stations were ever built alike, but this Memphis station is one of the more unusual. Featuring a lower and wider roof and massive chimneys, it is likely a later design and was probably brand new when this 1950 photograph was taken. *Courtesy of Chip Flohe*

This Memphis repair shop was selling Texaco gasoline as well as doing repair work when this photo was taken in 1950. *Courtesy of Chip Flohe*

In 1950 this auto parts store on Hollywood Avenue in Memphis was selling Texaco gasoline and fiercely competing with a discounter location, believed to be a Martin station, next door. *Courtesy of Chip Flohe*

Local private brand stations operated in many cities in competition with the major brand stations and the trackside discounters. This station on Chapman Highway in Knoxville was branded "Dixie" when this photograph was taken in 1950. The company, which was not affiliated with the franchise Dixie Distributors operation branded several "Dixie" stations in Knoxville until about 1954. *Courtesy of Chip Flohe*

101

Twin signs from a single mast-arm pole were commonly used at corner service station locations such as this 1950 Gulf station in Raleigh, Tennessee. *Courtesy of Chip Flohe*

In the postwar era, many former trackside operators built stations on "feeder routes," commuter routes into a city, to take advantage of the commuters' attraction to their discount pricing. This cottage-style Spur station on Chapman Highway in Knoxville was busy when this 1950 photo was taken. *Courtesy of Chip Flohe*

Sinclair had introduced modern station designs by the time this photo was taken in 1950 in Knoxville. *Courtesy of Chip Flohe*

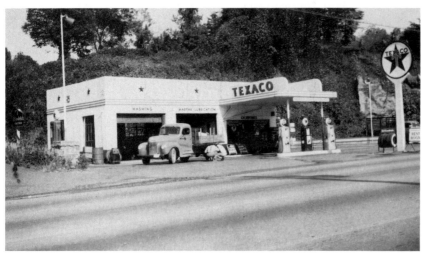

Considered the "most perfect industrial product package ever created," this classic Texaco station was located on Knoxville's Chapman Highway in 1950. *Courtesy of Chip Flohe*

Displays of "Free Premiums," giveaways for multiple gasoline purchases verified by a coupon system, were common at discounter locations. A motorist could complete a set of dishes by repeated purchases from this Knoxville Spur station in 1950. *Courtesy of Chip Flohe*

There are numerous lines of symmetry and a certain visual perfection in the gas stations of yesterday, as is witnessed in this classic Shell station on Roan Street in Johnson City, Tennessee, in 1950. *Courtesy of Chip Flohe*

A spotlit porcelain Pure sign, complete with finned and fluted pole structure identifies this Pure "modern cottage," which still stands on North Roan Street in Johnson City, Tennessee. This photo was taken in 1950. *Courtesy of Chip Flohe*

An unusual natural color brick Esso station on Central Avenue in Kingsport, Tennessee, in 1950. The station was designed to harmonize with the appearance of nearby municipal structures. By the fifties, most brick stations had been covered with panels of porcelain enamel. *Courtesy of Chip Flohe*

While most major brand stations were involved in automotive repair by the postwar years, this tiny Gulf station in downtown Kingsport, Tennessee, appears to be operating a parking lot instead. *Courtesy of Chip Flohe*

A painted mural advertising Gulf products appears on the wall above this Gulf station in Kingsport, Tennessee, in 1950. *Courtesy of Chip Flohe*

Motorists entering Greeneville, Tennessee, on US 411 were greeted by this Shell station in 1950. *Courtesy of Chip Flohe*

This Cities Service station was giving S&H Green Stamps, as evidenced by the sign attached to the identification sign pole, when this photograph was taken about 1950. Discounters had long offered free premiums and stamp programs of their own, but a major brand gas station offering stamps from a national plan was a rarity at that time. *Courtesy of Chip Flohe*

This station in Erwin, Tennessee, in 1950 appears to have been a typical twenties structure with service bays added later. Note the ornate mast-arm pole supporting the Cities Service sign in this photo. *Courtesy of Chip Flohe*

This classic roadside country store was selling Gulf products at the time this 1950 photograph was taken in Tennessee. *Courtesy of Chip Flohe*

This Columbus, Montana, motor court was selling Mobilgas products when photographed about 1950. *Courtesy of Bob Mewes*

Visible pumps dispensing Esso gasoline were familiar sights at many country crossroad stores well into the fifties. This 1950 photo was taken outside Bowling Green, Virginia. *Courtesy of Thornton Beroth*

This roadside Esso station and garage on US 41A in Winchester, Tennessee, in 1950 featured rooftop mounted globes for identification. *Courtesy of Chip Flohe*

Wyllys Distributing was the Hudson Dealer in Alma, Michigan, and was offering "hometown" Leonard gasoline to the motoring public from Alma-based Leonard Refineries. *Courtesy of Chris McKee*

In the late thirties, Mexico nationalized its petroleum industry and "Pemex" became the only gasoline brand offered in Mexico. This postwar photo shows a stop at a canopied Pemex station about 1950. *Courtesy of Edwin Wharton*

Norfolk, Virginia-based Tankar Stations operated a series of gas stations constructed from actual railroad tank cars like this one on Hampton Boulevard in Norfolk in 1951. Tankar was one of the self-service pioneers, having converted to self service in 1949. *Courtesy Tankar Stations—Wayne Henderson Collection*

This series of photos represents a fortunate find for this book. The earliest photo, taken in 1950, shows a typical 1925 Esso dealer station in operation at the corner of Main and Love streets in Erwin, Tennessee. The second photo shows a new station under construction on the site. Look carefully at the earlier photo. Station construction had already begun (behind the old station) and these photos were taken only two weeks apart. The third photo shows the station in operation about 1952.
Courtesy of Chip Flohe

This station was operating as a Cities Service station and tire store when photographed about 1949. It is of unusual design having been built into a city block as opposed to a free-standing structure. By the time the 1952 photo was taken, the station had been converted into Direct Oil Company station, selling Spur products. *Courtesy of Chip Flohe*

The roadside service station, on the highway between "somewhere and somewhere else," is perhaps the most familiar scene of 20th century America. The station, typical of those built between 1925 and World War II, still stands along the road between Erwin, Tennessee, and Mars Hill, North Carolina. It is shown here as it appeared in 1952. *Courtesy of Chip Flohe*

A welcome addition to the neighborhood was a friendly service station. This Shell station was along the residential portion of Main Street in Erwin, Tennessee, in 1952. *Courtesy of Chip Flohe*

This Texaco station on South Roan Street in Johnson City, Tennessee, was participating in the "Registered Rest Room" program when photographed in 1952. *Courtesy of Chip Flohe*

Simplicity of design was the rule at this Memphis Shell station in 1952. Only the neon-outlined Shell sign identifies this station. *Courtesy of Chip Flohe*

Similar in design to some of the dignified Lion stations appearing in this book is this Memphis area Pure station in 1952. Tall Erie pumps and the ornately finned sign structure contribute to the dignified appearance. *Courtesy of Chip Flohe*

This twenties "sandbrick" Gulf station has been "modernized" in this 1952 photo with the addition of a carport-type shed over the outside grease pit. Both the forties center-mount pole and the postwar "Type K" pole support Gulf identification signs. *Courtesy of Chip Flohe*

This Cities Service station was under construction in downtown Johnson City, Tennessee, in 1952. *Courtesy of Chip Flohe*

The complex roofline contributes to the visual effect of this Pure Oil cottage station in Memphis in 1952. *Courtesy of Chip Flohe*

This classic Shell station was operating in downtown Springfield, Tennessee, in 1952. *Courtesy of Chip Flohe*

It's Christmastime at this Greenville, South Carolina, Atlantic station in 1952. Note the neon pylon-mounted Atlantic sign and the clock built into the front face of the pylon. This building design was the inspiration for a Bachmann "Plasticville" O-Scale toy train set service station in the mid-fifties. *Courtesy of Tony Lewis and R.V. Witherspoon*

This classic Gulf station is typical of the "federal" design Gulf used prior to the introduction of the porcelain "icebox" design in the late thirties. This station was built about 1936, and is shown here in a 1955 photograph. It was at the corner of Chelsea Avenue and Seventh Street in Memphis, Tennessee. *Courtesy of Chip Flohe*

This Memphis Mobilgas station was still using outside lifts when this photo was taken in 1954. When automotive services became commonplace at service stations in the early thirties, most stations built enclosed service areas. Apparently, that was not practical at this location. *Courtesy of Chip Flohe*

Chapter 6

1954-1965

The Creation of the Miracle Mile

The economic expansion that followed World War II was slowed somewhat by effects of the Korean War. The seeds of change had been sown, though, and as soon as the troops returned from Korea, the United States entered a period of economic growth unlike any seen before.

The automobile reigned supreme in this era, and cities and towns everywhere were reshaped to accommodate the mobile, automotive lifestyle. Suburban home building took people further and further from downtown. In what were formerly farm fields along highways between suburbia and

A neon Pure sign attracts the motorist passing through historic Jonesboro, Tennessee, in 1953. *Courtesy of Chip Flohe*

This unusual Texaco station in Mesa, Arizona, was selling Conoco motor oils along with Texaco products. Both the roof-mounted Texaco sign and the tiny Conoco triangle are unusual features. Also unusual is the barbershop in the sales office. Over the years *every* type of business has sold gasoline. *Courtesy of Bob Mewes*

downtown, businesses developed to serve the growing population and those new businesses were structured to automotive use.

Shopping centers, which were marketing experiments from before World War II, became a reality in most towns. Drive-in restaurants, drive-in theaters, and of course, service stations sprang up on the suburban scene. The sign industry, embracing the use of plastics, worked overtime to create the images necessary to attract attention in the soon-overcrowded strips. Indeed, the sign industry coined the term "miracle mile" for the suburban commercial areas.

In this chapter we see the gasoline marketers create new and larger stations and new images, highlighted by increased use of plastic signage.

Kansas City area jobber McCall Service was operating a chain of retail stations in that area in the early fifties. This photo was taken in Kansas City about 1953. *Courtesy of Jim Masson.*

Prefabricated steel service station structures were constructed from around 1910 well into the sixties. Shown here is a steel structure typical of those offered by Valentine Metal Buildings of Wichita, Kansas. It carries the brand of Kansas City Derby jobber McCall Stations and was photographed in 1953. *Courtesy of Jim Masson.*

By the early fifties some very successful service stations had expanded well beyond the normal two service bays. This Kingsport, Tennessee, Esso station is operating five service bays in 1953. *Courtesy of Chip Flohe*

Unusual-shaped lots created by intersecting city streets were ideal locations for gas stations such as this Shell station in Kingsport, Tennessee, in 1953. *Courtesy of Chip Flohe*

Standardized station designs were sometimes modified for existing sites, but this Esso station in Johnson City, Tennessee, in 1953 used a projecting office design usually reserved for narrow but deep lots found in downtown business districts. *Courtesy of Chip Flohe*

This unusual Pure station is constructed in a design more typically associated with Mobil, a large round corner area for sales office, although it does not appear to have been built as a Mobil station. It was located in Kingsport, Tennessee, about 1953. *Courtesy of Chip Flohe*

A gravel lot and above-ground tanks are typical of most trackside discounter gas stations. This station in Johnson City, Tennessee, was operated in 1953 by Nashville-based discounter Peoples Oil Company, which later sold out to Kerr McGee. *Courtesy of Chip Flohe*

An unusual center-mount pole supports the Gulf sign that identifies this Kingsport, Tennessee, Gulf station in 1953. *Courtesy of Chip Flohe*

A brand-new internally lit plastic sign identifies this Esso station in Kingsport, Tennessee, in 1953. *Courtesy of Chip Flohe*

"The Spot" diner and Pan-Am station formed an Elizabethton, Tennessee, landmark for many years. They're shown here in a 1953 photo. *Courtesy of Chip Flohe*

Both a thirties mast-arm sign pole and a fifties center-mount sign pole were used at this classic Sinclair station in Clarksville, Tennessee, in 1954. *Courtesy of Chip Flohe*

New Albany, Indiana-based Payless operated this station on Dickerson Road in Nashville in 1954. It appears to have recently converted from the "Dixie Dance" brand, a Payless predecessor. Payless would later pioneer the canopied pumper design that has been almost universally adopted by gasoline marketers nationwide. *Wayne Henderson Collection*

A revolving neon flying horse and a free-standing car wash (and you thought those were new!) are features of this huge Memphis Mobilgas station in September 1954. *Courtesy of Chip Flohe*

Although most gas stations were dealer-operated in the fifties and each had a "company name" identifying the dealer, it was a rare occasion when that name was displayed in any way more significant than lettering on the station facade or painted in the sales office window. Here the huge neon "Cotton States Servicenter" sign dwarfs the smaller Phillips 66 shields. This was a most unusual arrangement for 1954, when most oil companies forbid any identification signs other than their own at a station. *Courtesy of Chip Flohe*

Rocket gas pump globe attachments were used to promote the new DX Boron premium grade gasoline introduced in 1954. DX, Richfield of California, and Sohio would promote the addition of the chemical boron to their gasolines as a mileage enhancer. This station was located in Central City, Iowa, in 1954. *Courtesy of John Chance.*

This spectacular Speedway 79 station in Detroit, Michigan, featured a huge steel tower with neon outlining, topped by a huge neon "79." Pump globes are temporarily covered with a caricature of a ten-gallon hat for a special promotion. *Courtesy of Dick Doumanian*

The home and auto store was a familiar sight in cities and towns all across the country from the twenties until the onslaught of mass merchandisers in the seventies. This 1954 Memphis example sold Mobilgas products as well as Goodyear tires, appliances, and home and auto supplies. *Courtesy of Chip Flohe*

In "Miracle Mile" strips leading into cities, retail marketers had to compete for the attention of the passing motorist. Gasoline marketers had to transform their stations into attention-getting billboards, particularly at night, to compete on the crowded roadside. This 1954 Sohio station is an excellent example of eye-catching design. *Courtesy of Joe and Judy Gross.*

Ease of maintenance and timeless appearance were the primary reasons for porcelain panel construction. Shown here is an all-porcelain Memphis area Lion station from 1954.
Courtesy of Chip Flohe

In 1946 Mid-West Oil Company of Muskegon, Michigan, franchised the White Rose brand and logos from Cleveland-based National Refining. Gladwin Oil Company, the Gladwin, Michigan, White Rose jobber was operating this station, identified with both a modern plastic sign and the curbside "boy and slate," in 1954.
Courtesy of Chris McKee

This "jobber brand" McKinnon station was offering Conoco products when photographed in Memphis in 1954. *Courtesy of Chip Flohe*

X-L Oil Company was a Memphis discounter that operated several stations throughout Memphis in the postwar years. This station was located on Thomas Street in 1954. *Courtesy of Chip Flohe*

Another Memphis private brand was "Hanotex," an obvious takeoff on the famous Kanotex brand. This jobber brand location was apparently at one time a Pure station, judging from the hook-style mast-arm sign pole and the use of National A-38 pumps. This station was on Thomas Street in 1954. *Courtesy of Chip Flohe*

Gilbarco twin gas pumps are the most noticeable feature of this Memphis area Esso station in 1954. The twin pumps, one hose dispensing regular grade Esso and the other dispensing premium Esso Extra, made it easy for an attendant to "upgrade" the sale from regular to premium without the customer having to move his car. *Courtesy of Chip Flohe*

The search for "visually perfect" service station photos for this book required the use of these two photos. Both the Sinclair and the Mobilgas stations are of standardized designs repeated in town after town throughout the country, and they reinforce the idea that the service station was an "industrial package" for a range of products. Both stations are in Memphis in 1954. *Courtesy of Chip Flohe*

129

This classic Esso station was located in Memphis in 1954. *Courtesy of Chip Flohe*

This "Comoco" station is another example of a local discounter. It was located on Lamar Avenue in Memphis in 1954, and according to the signage, it was selling both Union Royal Triton and Conoco Super Motor Oils. *Courtesy of Chip Flohe*

Service stations like this Esso outlet in Memphis in 1954 were frequently located in industrial districts to appeal to fleet customers. Perhaps this dealer services vehicles for the lumberyard and feed mill in the background. It appears he could use *anyone's* business at the time this photo was taken! *Courtesy of Chip Flohe*

Neon rooftop letters are a postwar addition to this twenties-style Gulf station on Lamar Avenue in Memphis in 1954. *Courtesy of Chip Flohe*

Chelsea Oil Company was selling their "jobber brand" gasoline and Armstrong Tires from this location on Chelsea Avenue in Memphis in 1954. *Courtesy of Chip Flohe*

Automotive repair was an important part of any gas station's mix of business at the time this 1954 photo was taken. Note the porcelain "Tires" "Batteries" and "Accessories" signage on the station facade. *Courtesy of Chip Flohe*

Obviously tire sales were good business at this Memphis Shell station in 1954. Note the unusual pole for the larger, "second generation" Shell plastic light-up sign. *Courtesy of Chip Flohe*

Another "visually perfect" station was this Memphis Texaco location. *Courtesy of Chip Flohe*

When Shell Oil, in conjunction with plastics manufacturers and sign companies, developed the first internally lit plastic identification sign in the petroleum industry, a unique pole structure for the sign was also created. The side-mount pole is shown in this Memphis photo from 1954. *Courtesy of Chip Flohe*

This photo shows the bulk plant and flagship station belonging to Rex Oil Company in Thomasville, North Carolina, about 1954. *Courtesy of Rex Oil Company*

The red and white candy stripe design attracted the attention of the passing motorist to this Kayo station in Kingsport, Tennessee, about 1955. Adjacent to the Kayo was an unseen Pure station, in evidence by the porcelain Pure sign with neon arrows. *Courtesy of Chip Flohe*

The tiny Spur cottage design of the thirties had given way to a much larger cottage by the time this 1955 photo was taken on Knoxville's Western Avenue. *Courtesy of Chip Flohe*

Both Texaco gasoline and John Deere farm machinery were offered from this roadside enterprise in eastern Kentucky about 1955. *Courtesy of Chip Flohe*

This nondescript station offered Shell gasoline to a rural Appalachian community in east Tennessee in 1955. *Courtesy of Chip Flohe*

This roadside Sinclair station was located in Doeville, Tennessee, about 1955. *Courtesy of Chip Flohe*

This classic Shell station was located in Memphis about 1955. *Courtesy of Chip Flohe*

A typical scene from any small town main street in the fifties is the corner Sinclair station. This location was in Springfield, Tennessee, in 1955. Note the "S" over the bay doors, an unusual Sinclair design. *Courtesy of Chip Flohe*

Cottage designs for gas stations were not limited to use by the majors. LaFollette, Tennessee-based Fleet Oil Company built a number of these stations in Tennessee, Kentucky, and Virginia in the thirties and forties. This location is in Pineville, Kentucky, in 1955. *Wayne Henderson Collection*

Ashland Oil has long dominated the market in eastern Kentucky with stations like the one shown here in a 1955 photo. *Wayne Henderson Collection*

While most discounters located their stations in city and small town locations, this Kentucky station, branded by St. Louis-based Marine Oil, was a highway location with only a drive-in nearby. *Wayne Henderson Collection*

A transport fills underground tanks at a Chicago-area Oklahoma station around 1955. *Courtesy of Norm Oakes*

Bulk grade oils were offered by most of the major oil companies for their dealers to sell in 2gal refillable cans. This DX dealer has set up for an all-out promotion of Power Motor Oil, the DX bulk grade product. *Courtesy of John Chance.*

Oak Ridge, Tennessee, site of the development of the atomic bomb, was very much a boom town in 1955 when this classic "icebox" Gulf was photographed on Oak Ridge turnpike. *Courtesy of Chip Flohe*

This Chevron station and adjacent motel in Bonneville, Utah, was the fifties equivalent of today's truck stops and travel plazas. *Courtesy of Norm Oakes*

This roadside store in Topock, Arizona, offered Chevron gasoline among the product mix. *Courtesy of Patrick Shimmin*

This 1956 Pure station in Memphis, with billboard signage and no service bays, was obviously built for a discounter and later converted to Pure. Hundreds of trackside discounter locations were closed during World War II when gas shortages forced refiners to channel all of their output to branded stations, leaving no product available for independents to purchase. *Courtesy of Chip Flohe*

The Teague Texaco design was versatile because stations could be built with any number of service bays (or without any) and with or without a projecting canopy. This station in Memphis in 1956 has the projecting canopy with prominent Texaco lettering. *Courtesy of Chip Flohe*

This 1956 Knoxville Sinclair station is identified by a large rooftop neon Sinclair sign. *Courtesy of Chip Flohe*

In the thirties Gulf built a number of these art-deco "icebox" stations with round-end windows and a pylon tower with glass block and neon Gulf lettering. The location shown is in Memphis in 1956. *Courtesy of Chip Flohe*

This elaborate Pure Oil "cottage" station was located in the Germantown section of Memphis in 1956. *Courtesy of Chip Flohe*

While most gas stations were free-standing buildings and not a part of main street blocks, this Gulf station in Halls, Tennessee, was built into the main street business district. *Courtesy of Chip Flohe*

This classic Lion station was operating in Memphis in 1956. *Courtesy of Chip Flohe*

This Pan-Am station was operating in Memphis, Tennessee, in 1956. *Courtesy of Chip Flohe*

This Esso station, located on Bellevue Boulevard in Memphis in 1956, is typical of Esso company owned and dealer-operated locations from this era. *Courtesy of Chip Flohe*

This classic Pan-Am station was located in Memphis, Tennessee, in 1956. *Courtesy of Chip Flohe*

This cottage-style station, branded Cities Service, actually appears to have been built as a Pure Oil station. Numerous marketers, Pure, Cities Service, Phillips 66, Conoco, Skelly, Fleet, Spur, and others used variations of "cottage" design. *Courtesy of Chip Flohe*

This seemingly typical icebox Gulf in Memphis in 1956 features an unusual neon-outlined Gulf identification sign. *Courtesy of Chip Flohe*

Central Avenue in Memphis was the scene of this classic Gulf station in 1956. *Courtesy of Chip Flohe*

The corner service station was very much a part of every suburban neighborhood by the time this photo was taken of a Pan-Am station in Memphis in 1956. *Courtesy of Chip Flohe*

There was often a choice of four brands of gasoline at each intersection in the fifties. Competing here are a classic Sinclair station and an unseen Cities Service. *Courtesy of Chip Flohe*

This classic independent or "local" discounter was branded "X-L." This station was located at 4795 Poplar Ave. in Memphis in 1956. *Courtesy of Chip Flohe*

In 1957 Shell became the first gasoline marketer to introduce a rancher-style station, primarily in response to requests for neighborhood aesthetics. This 1956 Memphis-area station contrasts greatly with the rancher designs that would be introduced only a year later. Note the neon Shell sign still in use. *Courtesy of Chip Flohe*

Even in the days of gas stations on every corner it was unusual to see stations of the same brand competing at an intersection. This photo shows both a country store selling Gulf gasoline and a two-bay Gulf dealer station, a most unusual combination for the Memphis area in 1956. *Courtesy of Chip Flohe*

Towering island lights and the 1947-style plastic Shell sign, the first internally lit plastic gasoline sign, are featured prominently at this Memphis Shell station in 1956. *Courtesy of Chip Flohe*

Hesselbein Tire and Oil was a Memphis-area Firestone dealer and offered gasoline under its own "Hesselbein" brand. Their station is shown here in a 1956 photo. *Courtesy of Chip Flohe*

This Memphis Esso truck stop featured an upstairs lounge area so weary truck drivers could rest while repairs were being made. Superpremium Golden Esso Extra is among the products offered here in 1956. *Courtesy of Chip Flohe*

One of the earliest internally lit plastic gasoline signs was the Esso oval introduced about 1950. This typical Memphis station is identified by one of those early signs in this 1956 photo. *Courtesy of Chip Flohe*

The Pure "modern cottage" station design was introduced in the late forties to replace the familiar "Pure cottage" design. This location in Memphis in 1956 is typical of the design in its earliest form. Later a pylon chimney and rooftop signage was added. *Courtesy of Chip Flohe*

Neon rooftop signage identifies this Memphis-area Esso station in 1956. Also featured in this photo is the early plastic internally lit Esso identification sign. *Courtesy of Chip Flohe*

One-stop shopping! This suburban Memphis Esso station was also a drug store, diner, and, of course, did automotive repair. Although this photo dates from 1956, this station is still displaying the oval "Standard Esso Dealer" sign that was discontinued shortly after World War II. *Courtesy of Chip Flohe*

This 1956 Memphis Gulf station was one of the prototypes for the sixties Gulf station design. The station featured large expanses of glass and an internally lit plastic Gulf old disc sign. *Courtesy of Chip Flohe*

The Stanton-Barbee Mobilgas station was located on Poplar Pike in Memphis when this photo was taken in 1956. *Courtesy of Chip Flohe*

Of every style of commercial or industrial building ever designed to be constructed in mass quantities, the classic Texaco station that industrial designer Walter Darwin Teague created in 1937 is perhaps the most widely used. Many variations were created, but in more than twenty years of active construction, well over 10,000 were built—or created by modifying existing buildings. *Courtesy of Chip Flohe*

This auto repair garage at Fort and West Grand in Detroit featured Speedway 79 products, complete with a Speedway 79 neon sign from the thirties, when photographed here in 1956. *Courtesy of Dick Doumanian*

This Leonard station in Michigan was selling DX Motor Oils as well as Leonard Gasoline when the 1952 photo was taken. Siberling tires were added to the product lineup by the time the 1956 photo was taken. *Courtesy of Chris McKee*

Phillips Petroleum's postwar building boom featured a station design with canted show windows and simulated rock around the office area, as is shown in this 1957 Memphis photo. *Courtesy of Chip Flohe*

In 1957 the Pan-Am brand was replaced by Amoco. This Memphis station was displaying Amoco pump signs and a "Distributor of Amoco Products" sign attached to the Pan-Am identification sign, in preparation for the conversion to the Amoco brand. *Courtesy of Chip Flohe*

This was a typical fifties Pure Oil station, located in Memphis, Tennessee. *Courtesy of Chip Flohe*

The Memphis suburb of Raleigh, Tennessee, was home to the classic Gulf station shown in this 1957 photo. *Courtesy of Chip Flohe*

This Memphis auto repair shop was offering Sinclair gasoline in 1957. *Courtesy of Chip Flohe*

Only the newer style pumps and vintage 1936 Gulf sign and pole have been added to this otherwise intact twenties-style Gulf sand brick station, located in Memphis. This photo was taken in 1957. *Courtesy of Chip Flohe*

Contrast the 1957 modern "icebox" Gulf, a three-bay shopping center location, with the simple sand brick Gulf station shown in the previous photo. Both were operating in Memphis in 1957. *Courtesy of Chip Flohe*

Truck stops, an invention of Toledo-based Pure Oil affiliate Hickock Oil (Hi-Speed), were experimented with by several other marketers in the postwar years. Seen here is a Pan-Am truck stop in Tennessee, a design they standardized and would repeat throughout the south in the fifties. *Courtesy of Chip Flohe*

In contrast to the standardized design of the Pan-Am truck stop is this nearby Gulf-branded outlet. Anyone fortunate enough to operate a cafe with a large lot along a major truck route could simply add gas pumps and become a "truck stop" overnight. Gulf experimented with assigning their truck stops numbers and issuing maps directing truckers to their locations. *Courtesy of Chip Flohe*

151

Cutler Oil Company, a Leonard jobber in Lansing, Michigan, supplied this location in Lansing. In the 1950 photo the station was serving as a Packard dealership, but by the time of the 1957 photo, the Packards were long gone. *Courtesy of Chris McKee*

This station in Edgerton, Wisconsin, was offering River States gasoline when photographed here about 1955. River States became RS Royal when Murphy Oil purchased the company in 1957. After Murphy purchased Spur Distributing in 1960, marketing was consolidated under the Spur brand. The extensive River States/Royal dealer network in Wisconsin and Minnesota would bring the familiar southern brand Spur into entirely new territory, but obviously it was successful as it continues today. *Courtesy of Larry Witzel*

The Edgerton, Wisconsin, station after conversion to Spur, about 1963.
Courtesy of Larry Witzel

Motorists traveling California's US 101 could purchase Richfield gasoline at dozens of eagle-topped stations like this one in Los Angeles in June 1958.
Courtesy of Walt Wimer, Jr

One of the most ornate "castle" stations ever constructed is this Monroe, Michigan, Leonard station, shown here in a 1958 photograph. *Courtesy of Chris McKee*

Rooftop signage was commonly used in "California" design stations of all brands. This classic 1958 Seaside station in Morro Bay, California, sports a rooftop sign as well as neon identification signs. *Courtesy of Walt Wimer, Jr*

Go Signal! Standard of California's affiliate brand Signal marketed products all along the Pacific Coast through stations like the one in La Jolla in 1958 shown here. Note the "boulevard sign," another California station feature. *Courtesy of Walt Wimer, Jr*

Walt Wimer's 1957 Chevy was being filled by a uniformed attendant while he was preserving, for all to enjoy, this scene in Hope, British Columbia, in June 1958. *Courtesy of Walt Wimer, Jr*

Roadside cafes often sold gasoline as well as food very much like today's travel plazas and convenience stores. Shown here is a combination diner and Lion gas station from 1959. *Courtesy of Chip Flohe*

Henley Street in downtown Knoxville was the sight of this typical "icebox" Gulf station, shown here as it appeared around 1959. *Courtesy of Chip Flohe*

The Spur Distributing Company opened its sixth station, at Broadway and Depot, in Knoxville in 1928. Here it is shown in a 1959 photo, featuring the rare Spur "Tank Car" globes on Tokheim Model 39 pumps. Shortly after this photo was taken, Spur Distributing was sold to Murphy Oil. *Courtesy of Chip Flohe*

The open road awaits the motorists filling up at this Vickers station in Rexford, Kansas, in 1959. Look for the 1957 Chevy convertible in other scenes in this book, as this photo is part of a series taken by Walt Wimer, Jr. beginning in 1958. *Courtesy of Walt Wimer, Jr*

In 1962 the former Spur Distributing Company stations throughout the south were reimaged with the new Murphy Oil/Spur signage and colors. This photo shows Spur #6, at Broadway and Depot in Knoxville in November 1963. *Courtesy of Chip Flohe*

Futuristic "Batwing" design was certainly an attention-getter at this Wichita, Kansas, Vickers station in August 1959. *Courtesy of Walt Wimer, Jr*

Another classic "Batwing" Vickers station in the Wichita, Kansas, area in August 1959. *Courtesy of Walt Wimer, Jr*

Today's gasoline-oriented convenience store was born right here. Picnic supplies and sporting goods were offered from this crowded Minneapolis, Minnesota, Erickson station in August 1959. With each purchase you qualified to receive Erickson's "Holiday Savings Stamps," which were good toward various free premiums displayed in glass cases on the pump island. By the following year the Holiday name and rocket logo was applied to a larger convenience store concept—the Holiday Stationstore. *Courtesy of Walt Wimer, Jr*

Another Minneapolis-area Erickson station begins the conversion into a convenience store in August 1959. *Courtesy of Walt Wimer, Jr*

Hello Neighbor! Pate stations, including this one in suburban Milwaukee in 1959, offered this greeting in neon on stations throughout southeastern Wisconsin. This station shows both the old Pate shield and the new Pate oval, introduced when Standard of New Jersey (Esso) purchased Pate in 1956. Enco would replace the Pate brand only two years after this photo was taken. *Courtesy of Walt Wimer, Jr*

Knight Oil Company was operating this garage and service station in the company's hometown of Springfield, Missouri, in August 1959. Gasoline sales were handled from the small building at the corner of the lot while repairs were being made in the garage building. *Courtesy of Walt Wimer, Jr*

A twin pump and clear ripple pump globes were unusual features at this Crystal Flash station in Grand Rapids, Michigan, in August 1959. *Courtesy of Walt Wimer, Jr*

The huge "Better Buy Bay" sign dwarfs the Bay station at this Denver crossroads in August 1959. *Courtesy of Walt Wimer, Jr*

A rooftop billboard advertises the new "Hottest Brand Going" advertising campaign at this Denver, Colorado, Conoco station in August 1959. *Courtesy of Walt Wimer, Jr*

This Kingfisher, Oklahoma, DX station was a showplace of the new DX color scheme when photographed here in August, 1959. *Courtesy of Walt Wimer, Jr*

Like an "enchanted cottage" from a children's storybook, this classic Phillips cottage was still displaying the pre-1959 Phillips logo and images when photographed outside Oklahoma City in August, 1959. *Courtesy of Walt Wimer, Jr*

This Speedway 79 station design was introduced in 1959 and featured an illuminated facade constructed entirely of plastics. The design was adopted by Marathon station design after Marathon purchased Speedway in 1959 and hundreds were built with the Marathon brand. *Courtesy of Dick Doumanian*

Every Southern discounter started out as just a local marketer, much like Dave's Oil Company of Greeneville, Tennessee. Shown here is Dave's Newport, Tennessee, station in 1960 with the large above-ground tanks painted to read "Save With Dave." *Courtesy of Chip Flohe*

Superpremium Gulf Crest was the featured product at this timeless Gulf station on Summer Avenue in Greeneville, Tennessee, in 1960. *Courtesy of Chip Flohe*

In 1937 Texaco commissioned prominent industrial designer Walter Darwin Teague to create the "perfect" service station design. Shown here is an excellent example of what became known as the "Teague Texaco." This station was located between Kingsport and Johnson City, Tennessee, and was photographed in the winter of 1960. *Courtesy of Chip Flohe*

This typical "neighborhood" Gulf station was located in Oak Ridge, Tennessee, in 1960 as part of a mini shopping center. *Courtesy of Chip Flohe*

In 1959 Phillips Petroleum introduced its new logo and reimaged its stations. This February 1960 photo shows the new image in place, with the old logo still displayed on shield-shaped gas pump globes. *Courtesy of Chip Flohe*

Thomasville, North Carolina, Crown jobber Rex Oil Company had private-branded this former Crown location to its own Rex brand when this photo was taken in Thomasville in 1960. *Courtesy of Rex Oil Company*

The Rex Oil Company station in Thomasville, North Carolina, had been remodeled and Crown had replaced Crown and Crownzol with Crown Gold and Crown Silver by the time this 1960 photo was taken. *Courtesy of Rex Oil Company*

This Oak Grove, North Carolina, country store was selling Crown gasoline at the time this 1960 photograph was taken. Note the large three-dimensional Coca Cola bottle by the door. *Courtesy of Rex Oil Company*

Crown, with its elaborate logo, was an early convert to plastic station signage. This Thomasville, North Carolina, station featured an internally lit plastic Crown sign in 1960. *Courtesy of Rex Oil Company*

The island kiosk was not an invention of the seventies' self-serve pumpers. This Imperial station on South Woodlawn Boulevard in DeLand, Florida, in 1960 used this booth on the pump island to protect attendants from the elements. *Courtesy of Walt Wimer, Jr*

Crown gasoline was available from this Unity, North Carolina, station in 1960. *Courtesy of Rex Oil Company*

Colonial "Minuteman" products were offered at this station high above US 17/92 in Orange City, Florida, in 1960. *Courtesy of Walt Wimer, Jr*

Winston-Salem-based Travelers Oil had not yet removed the metal band globes from the gas pumps at this Fayetteville, North Carolina, station when it was photographed in March 1960. *Courtesy of Walt Wimer, Jr*

This Phillips station design is rarely seen with the old Phillips logo and image. This rare photograph was taken in Orlando, Florida, in 1960. *Courtesy of Walt Wimer, Jr*

Taylor Oil Company's Etna brand was being introduced at some former Travelers stations in North Carolina when these two stations were photographed in Greensboro in March 1960. Both the Travelers and Etna brands are still in use today, although by separate companies. *Courtesy of Walt Wimer, Jr*

This tiny Red Ace station in Nashville in 1960 resembled the station designs used by both Shell and Sinclair with the pointed pylon above a canted roofline. *Courtesy of Walt Wimer, Jr*

The dozens of blinking lights on the sign and pole structure was sure to attract attention to this Nashville area Ingram station in 1960. Less than a year later, the Spur brand would begin to replace Ingram throughout the south. *Courtesy of Walt Wimer, Jr*

The distinctive Leonard sign and pole structure identifies this porcelain paneled Leonard station in Shield, Michigan, in 1960. *Courtesy of Chris McKee*

Hudson stations, the creation of Mary Hudson of Kansas City, Kansas, dotted the landscape from coast-to-coast. While most discounters preferred to operate in a limited geographical area, Hudson stations were common sights in cities and towns all over the country. This station, typical of many, was photographed about 1960. *Courtesy of Jim Masson.*

An open-air "carport" type service area was still in use at this Vero Beach, Florida, Pure station in 1961. *Courtesy of Walt Wimer, Jr*

The chimney-like pylon, featuring an internally lit Shell logo, was the focal point of the fifties Shell station design like the one shown here from east Tennessee in 1961. *Courtesy of Chip Flohe*

Twenty-five years after the logo was replaced, this Pierson, Florida, roadhouse was still selling Texaco products under the early Texaco "Gasoline/Motor Oil" sign in 1961. *Courtesy of Walt Wimer, Jr*

This attention-getting "windmill" station sold Texaco products for many years in Ohio. *Scott Benjamin Collection*

Hess Oil and Chemical entered gasoline marketing in the late fifties with stations that were most unusual for that era. Constructed of aluminum-framed glass against a block wall and topped by an illuminated plastic header, the Hess stations were unlike anything else in the marketplace. This location was photographed in Pennsauken, New Jersey, in 1962. *Wayne Henderson Collection*

What a mismatched station. Built as a Pure Oil cottage in Northern Florida in the thirties, by 1962 they were selling Texaco products under a barn roof canopy, with no less than two Texaco banjo signs. *Courtesy of Walt Wimer, Jr*

Although this Gulf dealer and this Spur station were fierce competitors, Spur had been supplied with "unbranded" gasoline almost exclusively by Gulf for over 30 years by the time this 1962 photo was taken in Cythiana, Kentucky. Unusual camera angle adds merit to this photo. *Courtesy of Chip Flohe*

The 1960 redesign of Shell signage featured the well-known logo on a red background as is seen in this Kingsport, Tennessee, station photographed from 1962. *Courtesy of Chip Flohe*

Sinclair stations such as this one were familiar scenes all over the country until the 1970 merger with Atlantic Richfield. This location in Kingsport, Tennessee, shown here in 1962, survived to rebrand Arco in the seventies before closing around 1980. *Courtesy of Chip Flohe*

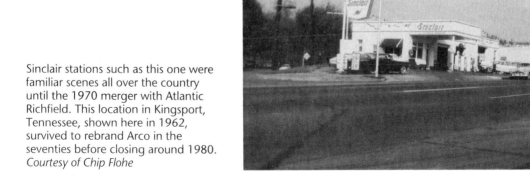

Here a lakeside bait and tackle shop offers Sinclair gasoline to passing motorists between Kingsport and Johnson City, Tennessee. It was still displaying the old "HC" logo when photographed in 1962. *Courtesy of Chip Flohe*

In 1957 Pan-Am stations in the southeast were rebranded Amoco. Then, in 1961 Amoco stations in the east and south were converted to American, much like the one shown in this 1962 photo from Greeneville, Tennessee. *Courtesy of Chip Flohe*

Replacement American faces were quickly installed in existing Pan-Am/Amoco sign frames as a temporary measure when the American brand replaced Amoco in the south in 1961. Shown here is a Greeneville, Tennessee, station with the temporary sign in 1962. *Courtesy of Chip Flohe*

The Esso brand appeared on many large and elaborate stations like this one at the traffic circle in Kingsport, Tennessee, in 1962. By the time this photo was taken, the "Humble" corporate name had been added to the station building. *Courtesy of Chip Flohe*

Many bulk plant operators built their "flagship" stations adjacent to their plants as is shown in this 1962 photo of a Kingsport, Tennessee, Esso station with bulk storage tanks in the background. *Courtesy of Chip Flohe*

Always a favorite of children was the candy-striped Kayo stations throughout the south. This station with striped sign poles and building facade was located in Greeneville, Tennessee, in 1962. *Courtesy of Chip Flohe*

Greenwood, Mississippi-based discounter Billups Brothers had became the Billups Western Division of Signal Oil and Gas by the time this 1962 photo was taken in Hernando, Mississippi. *Courtesy of Chip Flohe*

Over the years virtually every other type of business has been combined with gas stations. In this photo, offices for a Hernando, Mississippi, mobile home park were located in this 1962 Standard Oil (Kentucky) station. *Courtesy of Chip Flohe*

Prominent on Main Street in every southern town was a neon Standard Oil sign as is shown here in front of the "Downtown Station" in Hernando, Mississippi in 1962. *Courtesy of Chip Flohe*

Ashland's "A-Plus" design, with large red "A" on the station side and "Flying Octanes" sign with arrows, is shown in this 1963 Louisville photo.
Courtesy of Chip Flohe

This Richmond, Virginia, Richfield station was still displaying a neon sign with the Richfield logo of the thirties when photographed here in August 1963. Less than a year later the Sinclair brand would replace the Richfield name at stations all along the east coast, as Sinclair had owned Richfield (New York) since the thirties, operating it as a secondary brand.
Courtesy of Walt Wimer, Jr

When Standard of California (Chevron) purchased Standard of Kentucky in 1961, long-standing supply contracts with Standard of New Jersey were canceled. As a result, Standard of New Jersey, through the domestic marketing subsidiary Humble Oil, entered Standard of Kentucky's former marketing territory, building hundreds of Esso stations in just a couple of years. Restrictions on the use of the Esso brand forced them to change the stations in those five states to Enco in 1967. This Louisville station displays the Esso brand in this 1963 photo. *Courtesy of Chip Flohe*

Perhaps the car speeding through this scene had just filled up with Crown Gold gasoline! This station was serving Richmond, Virginia, motorists when photographed in August, 1963. *Courtesy of Walt Wimer, Jr*

This Standard of Kentucky station in Louisville in 1963 has gas pumps labeled Chevron and Chevron Supreme. Gasoline brands were changed from Crown and Crown Extra to Chevron and Chevron Supreme when Standard of California purchased Standard of Kentucky in 1961. *Courtesy of Chip Flohe*

The classic "Teague Texaco," featuring a not-so-typical side-mount Texaco sign. While the standardized Texaco "banjo" sign was used for most locations, the side-mount pole was developed for locations where sign visibility was a problem, such as a main street setting or a cluttered business strip. This station was located in Memphis in 1964. *Courtesy of Chip Flohe*

In 1960 The Pure Oil Company introduced the station design known as the "Big One." Glass front, multiple service bays, and rooftop signage were features in this Memphis "Big One" in 1965. *Courtesy of Chip Flohe*

In the early thirties famed architect Frank Lloyd Wright designed a single service station for Phillips 66 in Cloquet, Minnesota. Within the overall reimaging of the Phillips 66 brand in 1959 was the introduction of the gas station design shown in this 1965 Knoxville photo. In this design, which Phillips continued to use until the early eighties, were numerous elements of the Cloquet station. Although Frank Lloyd Wright did not actually create this design, it is commonly attributed to him because of the numerous elements copied from his design. *Courtesy of Chip Flohe*

By the early sixties, the crowded Erickson's Holiday stations had evolved into this attractive Holiday "Stationstore" design concept. Erickson's was the first oil company to develop the convenience store concept that is now the industry standard. *Courtesy Holiday Stations.*

Chapter 7

1965-1974

Interstates and 'Keep America Beautiful'

While the "miracle mile" commercial strips thrived in the sixties, traditional highway gas stations met a drastic change. The Interstate highway system, a creation of the Eisenhower administration, made thousands of miles of two-lane highways obsolete overnight, ending the life of gas stations that had served travelers on those roads for decades.

Oftentimes rural highway stations could survive several years after being bypassed, but the massive gas station construction program launched by the oil industry proved to be the final blow. Oil companies built new stations at interstate interchanges in record numbers. High-rise identification signs, many with new trademarks developed for high visibility, sprang up alongside this new breed of station.

Another presidential program, the Johnson administration's "Keep America Beautiful" campaign, targeted gas stations and gas station signage for a high-visibility clean up. Rancher-style stations, brick-faced structures with residential rooflines, replaced the traditional porcelain-box stations. Size, location, and images all gave way in this era. In this chapter we'll see the changes.

Esso reimaged their stations in 1965, with narrow red stripes on white porcelain being replaced by large red bands on white as is shown in this 1965 photo of a Memphis station. *Courtesy of Chip Flohe*

185

The "Humble" name was added to Esso stations in 1961 and pump colors were revised in 1965 as shown in this 1965 Tennessee photo. *Courtesy of Chip Flohe*

When color televisions first became available, shopping for one in Memphis may have meant a trip to the corner Pure Oil station! Pure dealer Frank Black was operating this home and auto store in Memphis around 1965. One of the signs posted near the curb is a Motorola TV sign. *Courtesy of Chip Flohe*

The sixties saw the demise of many familiar gasoline brands as is shown in this Memphis Cities Service station converting to Citgo in 1966. *Courtesy of Chip Flohe*

The rooftop pegasus was still prominently featured in this sixties Mobil design as seen in this 1966 photo of a Memphis station. *Courtesy of Chip Flohe*

Aggressive gasoline discounters often located adjacent to successful major-brand stations in order to feature their discount price advantage. This Site station was located across from the Central Avenue (Memphis) Gulf station shown in chapter 6. *Courtesy of Chip Flohe*

St. Louis-based discounter Site displays the motto "Gas For Less" on their identification signs at this Memphis station in 1966. *Courtesy of Chip Flohe*

Improvements in dispenser technology are shown in the unusual gas pumps at this sixties Memphis Phillips 66 station. *Courtesy of Chip Flohe*

The 1959 Phillips 66 reimaging is quite attractive on this former Pan-Am station in Elizabethton, Tennessee, as photographed in 1966. *Courtesy of Chip Flohe*

This "modern" Sinclair station was located at the main intersection in Elizabethton, Tennessee, in 1966. *Courtesy of Chip Flohe*

Pure's station design had evolved from variations of the cottage to a simple porcelain box by the time this station was built around 1960. This station shown in this 1966 photo was located in Kingston, Tennessee.
Courtesy of Chip Flohe

This classic DX station was photographed through a car window while the driver was passing through Kingston, Tennessee, in 1966.
Courtesy of Chip Flohe

By the sixties the Teague Texaco design had been revised, and one such revision is shown in this 1966 photo taken in Kingston, Tennessee.
Courtesy of Chip Flohe

This Memphis Esso station had many unusual elements, including a flat canopy instead of the typical canted canopy and a vertical neon modular sign. *Courtesy of Chip Flohe*

As an extension of the 1960 station design that Pure Oil referred to as the "Big One," many major cities were the site of the even larger "Pure Car Care Centers," complete service facilities with up to twelve bays. Shown here is the Knoxville center in 1966. *Courtesy of Chip Flohe*

Liberty Oil Company, the Mt. Vernon, Illinois, discount marketer, operated this station on North Cleveland in Memphis in 1966. Typical of most discounters of that era, this location features multiple pump islands. *Courtesy of Chip Flohe*

Around 1966, the Rex Oil Company rebuilt the downtown Thomasville, North Carolina, station shown in chapter 6 and branded it Crown with the new "controlled background" logo. *Courtesy of Rex Oil Company*

In 1968 Union Oil signage was added to Pure station facades in anticipation of the eventual rebranding to Union in 1970. This station on US 17 at Tabb, Virginia, features rooftop neon Pure lettering above the Union logo. *Wayne Henderson Collection*

The domed glass Pure sign from the thirties was still a prominent feature of this roadside Pure station in Beech Grove, Tennessee, in 1968. *Courtesy of Chip Flohe*

Humble Oil, the domestic marketing subsidiary of Standard Oil of New Jersey, built hundreds of these rancher-style stations at interstate interchanges in the sixties. Many, like this Kingston, Tennessee, station were branded Esso, but the design was also used in Enco and Humble marketing territory as well. *Courtesy of Chip Flohe*

Interstate stations featured high-rise signage wherever possible, as is shown in this photo of an Oak Ridge, Tennessee, Texaco station with Texaco and Chevron high-rise signage in the background. *Courtesy of Chip Flohe*

In 1962 Standard Oil of California purchased Standard Oil of Kentucky with their Standard Oil-branded stations in Kentucky, Georgia, Florida, Alabama, and Mississippi. When stations were built in Tennessee in the sixties however, Esso (Standard of New Jersey) owned the rights to the "Standard" brand in Tennessee and stations had to be branded "Chevron." Long used on the West Coast, Chevron would eventually (1977) replace the Standard Oil brand in the south. This station was in Kingston, Tennessee, in 1968. *Courtesy of Chip Flohe*

In 1962 Gulf Oil and Holiday Inn reached an agreement where Holiday Inn would accept Gulf credit cards and Gulf would build stations at each Holiday Inn. Shown in this 1970 photograph is the Holiday Inn and Gulf station on I-40 at Oak Ridge, Tennessee. *Courtesy of Chip Flohe*

Crown was in the process of phasing out jobber contracts in favor of company direct marketing when Mobil re-entered North Carolina markets. Rex Oil Company, which had been a branded Mobil jobber in the forties, rebranded Mobil in 1967. The station shown in earlier in this chapter appears here as a Mobil station in this 1971 photo. *Courtesy of Rex Oil Company*

By the time this photograph was taken in 1971, a new facade had been added to this small station in Orlando, Florida, and a canopy covered the multiple-pump islands. Clayton, Missouri-based Imperial Refineries operated hundreds of stations like this one throughout the Midwest and in Florida. *Courtesy of Walt Wimer, Jr*

Flying A was in the process of rebranding to Getty when this station was photographed in Auburn, New York, in 1971. *Courtesy of Walt Wimer, Jr*

Discounter operations, such as this Hudson station in Burlington, North Carolina, in 1971, prominently displayed their attractive pricing. Most tried to stay a penny or two under the prevailing price for major brand gasoline. Product availability during the upcoming gas shortage would soon change their strategy to one of simply surviving the shortage. *Courtesy of Walt Wimer, Jr*

The end of the classic era of the American gas station is clearly marked by a single event and hence, by the single sign shown in this photo. This Esso station in Loudoun, Tennessee, in October 1972 has the Esso tiger holding an Exxon sign with the banner "We're changing our name." By early 1973 Esso was gone, the gas shortage was apparent, and gasoline marketing would never again be the same. *Courtesy of Chip Flohe*

Chapter 8

1974 -1985

Shortages, Self-Serve, and C-Stores

The 1973-1974 gas shortage changed gasoline marketing forever. Long lines, half-price pumps (many gas pump computers could only figure to 49 cents so the total price had to be doubled on sales to 99 cents), and self service were the order of the day.

When the dust settled from the first shortage, thousands of stations had closed. The second

Known as the "blue brick" design (side walls were constructed of blue brick), this station is the Marathon adaptation of the 1959 Speedway 79 design shown in chapter 6. This location is in Pontiac, Michigan, around 1974. *Courtesy of Dick Doumanian*

By the late sixties the "blue brick" station was beginning to look somewhat out-of-date, so Marathon converted to this colonial design. This location was photographed in Pontiac, Michigan, around 1974. *Courtesy of Dick Doumanian*

shortage of 1978 was somewhat less drastic, but all in all, a significant downsizing of retail gasoline marketing took place. Major oil companies converted many dealer-operated stations to a "split-island" arrangement, while the discounters wholeheartedly embraced self-service operations.

Convenience store chains often added gasoline as a sideline, while several of the regional oil

Known as the "Kettering" design from its introduction in Kettering, Ohio, this Marathon station style served as an alternate to the colonial style stations. This location was photographed in Kettering, Ohio, in 1974. *Courtesy of Dick Doumanian*

The Speedway brand was dormant for several years after being replaced by Marathon in 1962. However, by the seventies, after the gas shortage closed hundreds of dealer operated stations, many sites were converted to gas-only pumpers like the one shown here. The Speedway brand was reintroduced for a company owned chain of pumper type stations. In later years Speedway would become the primary brand for Marathon's Emro marketing and it currently appears on hundreds of stations throughout the South and Midwest. *Courtesy of Dick Doumanian*

companies followed the lead that Erickson's had taken in 1959 and began to develop their own store concepts. Tenneco and Ashland were early store operators, Ashland having purchased the Erickson affiliate SuperAmerica.

Major oil company jobbers were quick to develop new stores, many having supplied neighborhood and country stores for most of this century. And by 1980, virtually all of the major oil companies were converting former dealer-operated, repair-oriented stations into convenience stores. In this chapter we'll take a quick look at this era of change in gasoline marketing.

One of the companies absorbed by Emro/Speedway was Chicago-based Cheker. This location is a converted Marathon dealer location in Grand Rapids, operating "gas-only" under the Cheker brand. *Courtesy of Dick Doumanian*

197

Environmental requirements of the seventies brought about the need for unleaded gasolines. While most marketers added a regular grade unleaded to their existing product mix of a leaded regular and a leaded premium, Clark Oil, long a single-grade marketer, introduced its unleaded as a premium grade, both for the new cars that required unleaded and for extra octane in the older cars that didn't. Clark was also among the last of the major brands to regularly use pump globes. This station was photographed in Anderson, Indiana, in 1975. *Courtesy of Walt Wimer, Jr*

Self service took a solid hold on the gasoline market after the 1973-1974 gas shortage. Traditional discounters, such as Savings Stations, operator of this station at Fletcher, North Carolina, would quickly adopt self-serve as a cost-cutting survival measure. *Wayne Henderson Collection*

Old-line discounters did little to enhance thier image, believing that the only attraction they needed was their posted prices. This Red Diamond station at Brevard, North Carolina, had changed little since the forties except for the addition of pumps that were console-controlled for self-service operations. *Wayne Henderson Collection*

Regional marketers, such as Tipton, Georgia-based Dixie Oil, took advantage of the confusion of the shortage years to upgrade their image and look like "mini-majors" within their marketing areas. Dixie is the last marketer regularly using gas pump globes on their pumps, as shown in this photo from Elberton, Georgia. *Wayne Henderson Collection*

Farm co-ops, long having supplied bulk fuels and lubricants in rural areas, expanded their direct marketing by adding self-service pumps at their feed stores, grain elevators, and other non-traditional outlets, while continuing to supply their own traditional branded outlets. This Co-op station is at Grainfield, Kansas. *Wayne Henderson Collection*

Clark stations were quick to adopt modern marketing techniques and combine them with their traditional image. This location is in Taylor, Michigan. *Wayne Henderson Collection*

Canopied "pumper" stations—multiple islands under a huge canopy with adjacent buildings for rest rooms and office space—became an industry standard for the seventies. Shown here is a Hess pumper at Anderson, South Carolina. *Wayne Henderson Collection*

The fueling activity at a pumper station was often controlled by cashiers in a central booth called a "kiosk." This East Coast station in the Manchester section of Richmond, Virginia, is typical of the pumper layout, with sixteen single pumps. It is built on the site of the first East Coast station on Hull Street, at one time Richmond's discounter row. *Wayne Henderson Collection*

Drive-through service. This Tops Oil Co station in Raleigh, North Carolina, uses a forced traffic pattern to in effect create a drive-thru gas station. Pump islands are approached from a main street from one entrance only and traffic is funneled through past a cashiers' booth—fill up and pay on your way out. Several marketers went one step further and actually had station personnel *living* in the cashier building, often a mobile home, and collecting payment from the comfort of their living room. *Wayne Henderson Collection*

Lighthouses, icebergs, windmills, airplanes—gas station design has taken many fantasy forms. Many gasoline marketers felt they had to do something out of the ordinary to catch the attention of the passing motorist. This "winged pyramid," long a Morristown, Tennessee, landmark, is a relatively late (1962) addition to the list of "fantasy" stations. *Wayne Henderson Collection*

Electronics and image were keys to service station design in the eighties. This Kerr McGee outlet in suburban Nashville (built on the site of the first Spur Distributing station) is an excellent example of how a marketer's image package, when combined with the latest in electronic equipment, presents an integrated image imminently identifiable with their brand. *Wayne Henderson Collection*

In Danville, Illinois, Wareco's classic "attendant" logo seems somewhat out of place in today's self-service world. The uniformed station attendant is now a part of days long gone. *Wayne Henderson Collection*

Chapter 9

1985-1994

Modern Times—'Images and Environmental Concerns'

The late eighties saw two factors change the face of gas stations, and change it in ways more drastic than any that had come before. Until this era, rural stations that could survive on forgotten highways were generally left as they were. Oil companies and jobbers had no reason to invest money in marginal stations, so they often retained images from decades past. Then, as oil companies downsized in this era, often returning to their original marketing areas, they looked to

In a sense, modern gasoline marketing began on this corner. At this crossroads in Barron, Wisconsin, the Erickson family, who own Holiday and created SuperAmerica, opened their first gasoline station in 1931.

Now the corner is occupied by a huge canopy and modern convenience store under the Holiday brand and image. *Wayne Henderson Collection*

Convenient multi-product dispensers and 24-hour operations, like those at this Hampton, Virginia, Getty station, are hallmarks of today's marketer. *Wayne Henderson Collection*

In other areas the traditional brands have been replaced with the new and unfamiliar. This Nashville BP station was undergoing the transition from Gulf to BP when photographed in 1989. *Wayne Henderson Collection*

New images are applied over the old. This Exxon station in downtown Sevierville, Tennessee, shows nineties graphics displayed over forties porcelain panels. *Wayne Henderson Collection*

make the most of their brand images. Marginal station operators had to either invest in canopies, lighting, signage, public rest rooms, and such, or lose their brand.

Most chose not to make the investment, and regional "jobber" brands appeared in large numbers. A typical jobber would have the oil company name and image on his direct-operated stations and the dealer-owned outlets that met the requirements, and his own private brand on other stations that could not meet image specifications.

In the midst of the image problem came environmental concerns. Several isolated instances of underground tank failure brought about an aware-

Here the image transformation even extend to the station signage. At this Norfolk, Virginia, Texaco station, the modern Texaco logo appears in a traditional banjo pole.

Sign and historical preservation ordinances sometimes create some unique station signage situations. *Wayne Henderson Collection*

In harmony with the surrounding countryside, even the dinosaur looks comfortable in this setting in Manitou Springs, Colorado. *Wayne Henderson Collection*

Canopy design and graphics have become an important part of each station's image. The unusual canopy arrangement at this Marion, Illinois, Coastal station carries the Coastal graphics well. *Wayne Henderson Collection*

There seems no limit to the size of either the store or the canopy, as is evident at this Columbia, South Carolina, Hess station. *Wayne Henderson Collection*

ness of the problem of ground water being contaminated by gasoline. In 1987 the U.S. Congress enacted tank upgrade and replacement requirements, to be met over a ten-year period. More marginal stations fell by the wayside as marketers could not justify the huge investment necessary to continue operating.

In this chapter we'll see examples of typical stations from this era, carrying oil company image requirements and meeting EPA regulations.

Many traditional discounters, such as Winston-Salem-based Wilco, have made the transition to convenience stores very well. *Wayne Henderson Collection*

Farm co-operatives have also made many market advances, including some of the first 24-hour, unattended "cardlock" outlets. This Shell Lake, Wisconsin, Cenex station features a complete convenience store and hardware store under one roof, with automotive repair and traditional agricultural products nearby, forming a complete marketing complex. *Wayne Henderson Collection*

Even older C-stores make the transition to the nineties very well. This Chippewa Falls, Wisconsin, SuperAmerica from the early sixties has adapted well to nineties marketing and image. *Wayne Henderson Collection*

Modern station design lends itself well to seasonal promotions, such as the Christmas display at this Rice Lake, Wisconsin, Spur Station on Christmas Eve 1993. *Wayne Henderson Collection*

Back to the future? This modern, canopied convenience store at Reidsville, North Carolina, operates under the Pure brand. Many traditional brands have survived as regional marketers. The Pure brand was reintroduced in the south when Unocal withdrew from branded marketing there. *Wayne Henderson Collection*

Chapter 10

Survivors

The stations shown in this chapter all have one thing in common: At the time they were photographed, they were relatively intact images of another era. Many have since been altered or removed, in the wake of image requirements and environmental laws.

Let's take a final, historical look back at gasoline marketing throughout this century in these few stations. All photos in this chapter are from the Wayne Henderson Collection.

This station stood alongside old NC Route 36 just south of Mars Hill, North Carolina, from 1929-1989. Over the years Esso, Sinclair, Shell, and finally Crown products were sold here. When photographed here in 1975, it had been branded Crown for over twenty years.

Classic cottage architecture on four levels, this Gulf station in Courtland, Virginia, was built in 1936. The station is located on US 58, the main highway from Virginia's "Tidewater" area to North Carolina. I've passed this station all my life and it's long been a favorite. In 1984 I stopped long enough for this photograph.

While most cities and towns outlawed curb pumps in the twenties, this survivor, a neighborhood store in historic West Point, Virginia, was still selling gas at curbside in 1985.

Phillips Petroleum's distinctive cottage architecture is still apparent in this 1990 photo of this enlarged version of the classic Phillips cottage in Lawrence, Kansas.

This photo is included for historical reference since it is the only gas station ever designed completely by famed architect Frank Lloyd Wright. Pylon signage, canted windows, and an industrial look influenced gas station design from the time of its construction in 1932 well into the seventies. It's shown here in a 1990 photo.

The modern Amoco image is adapted to these two survivors—a Standard "mission" station at Dearborn, Michigan, and an Amoco "porcelain palace" at Marion, North Carolina. Both were photographed in 1989.

The icon of the miracle mile—the Clark "Circle of Service" sign with orbiting lights—has, by 1989, been amended with signs of modern times, notably the credit card acceptance signs and the unleaded heading for the price sign.

You could still get attendant service of Ashland products, albeit on a split self/full service arrangement, at this Ashland station at Kermit, West Virginia, in 1993.

A pair of Conoco "cottages," a traditional design from the thirties, still serve their original purpose. In Richmond, Virginia, Conoco has marketed as Kayo since 1959, but in Jefferson City, Missouri, this station has displayed the Conoco brand for over 50 years.

Appalachia Gulf has served the needs of motorists in Appalachia, Virginia, since 1923. It is seen here in a 1985 photo.

A dealer's determination has kept the Deep Rock brand alive at this Chippewa Falls, Wisconsin, Deep Rock station. It was photographed in 1987.

The Deep Rock image has also survived at this fantastic station in Beloit, Wisconsin, as photographed in 1990.

New meets the old. Modern canopy design and the traditional Marathon "blue brick" design combine to create an appealing image at this Lincoln Park, Michigan, Marathon station in 1989.

They were still doing business under the Flying Red Horse at Upper Sandusky, Ohio, in 1987.

A classic Phillips image is preserved in Hendersonville, North Carolina, in 1986.

A timeless porcelain facade still gleams in the sun at Mosheim, Tennessee. This is a 1989 photo.

Standard marketers have all but abandoned the Standard name, except in isolated instances where they maintain a "Standard" station so as to retain the rights to use the brand. This suburban Atlanta station mixes the Standard name with the modern Chevron image in this 1990 photo.

This Louisa, Kentucky, Sunoco station adapts the traditional image well, even to the use of the Sunoco wall mural on an adjacent building in this 1989 photo.

Leaded glass mixes with the Texaco "2000" image at Gloucester, Virginia. This station was photographed in 1984.

More than fifty years after its inception, the classic Texaco station is as timeless as ever, as shown in this 1984 photo.

Several variations of the Texaco image were used, including the one shown in this classic station in Clinton, South Carolina, in 1987.

In 1987 you would find the "Teague" Texaco image, banjo sign, and curb pumps as well at this Cape Charles, Virginia, garage and filling station. Texaco products have been sold here since 1907.

A neon cottage? Yes, a *neon* cottage, still pumping away at Superior, Wisconsin, in 1985.

Index

American, 115
Amoco, 8, 29, 68, 72, 73, 74, 75, 82, 83, 84, 115, 149, 178, 214
Arco, 177
Ashland, 7, 136, 181, 215
Associated, 45, 53, 56, 87
Atlantic, 30, 58, 60, 113, 177

B-A, 44
Bay, 163
Billups, 180
Bowser, 12
BP, 205

Carter, 8
Cenex, 209
Cheker, 179
Chelsea Oil Company, 131
Chevron, 5, 8, 137, 138, 182, 183, 192, 220
Citgo, 186
Cities Service, 30, 96, 105, 110, 112, 141, 143
Clark, 6, 7, 9, 198, 200, 215
Co-op, 199
Coastal, 207
Colonial, 169
Conoco, 71, 119, 128, 130, 163, 216
Criteria, 60, 61, 62
Crown, 8, 167, 168, 169, 182, 183, 211
Crystal Flash, 162

Dave's Oil Company, 165
Deep Rock, 8, 217, 218
Derby, 120
Direct Oil Company, 117
Dixie Distributors, 5
Dixie Vim, 6
Dixie, 101, 199
DX, 125, 137, 148, 164, 189

East Coast, 201
El Camino, 21
Enco, 8, 182
Erickson's Holiday, 160, 184
Erickson's, 6, 7, 161
Esso, 6, 8, 20, 36, 58, 78, 92-95, 98, 104, 106, 107, 109, 121, 123, 129, 130, 141, 145, 146, 161, 179, 182, 185, 186, 190, 194, 211
Etna, 171
Exxon, 194, 206

Fina, 7
Fleet Oil Company, 135, 141

Gasamat, 8
Gay Oil Company, 25
Getty, 194, 205
Gilmore, 21, 46, 47, 50, 54, 57, 70
Gladwin Oil Company, 127
Gulf, 5, 28, 59, 65, 66, 73, 75, 76, 78, 80, 90, 91, 93, 99, 100, 102, 104, 106, 112, 115, 117, 118, 122, 130, 131, 137, 139, 140, 142, 144, 146, 150, 151, 157, 165, 166, 177, 187, 193, 212, 217

Hancock, 55
Hawkeye Oil, 14, 15, 16, 27
Hess Oil and Chemical, 76
Hess, 200, 208

Hesselbein, 144
Hickock Oil, 7, 151
Holiday, 7, 9
Hudson, 6, 7, 107, 174, 194
Humble Oil, 182, 192
Humble, 8, 186

Imperial Refineries, 193

K-T Oil Company, 17
Kanotex, 27, 128
Kayo, 133
Kendall, 39
Kerr McGee, 7, 8, 122, 202
Knight Oil Company, 162

Lamson Oil Company, 26
Lamson's Nun-Bet-Er, 24
Leonard, 81, 107, 148, 152, 155, 173
Liberty Oil Company, 190
Linco, 72
Lion Oil, 100, 127, 140, 157

Marathon, 5, 12, 72, 88, 165, 195, 196, 197, 218
Marine Oil, 136
Martin Oil Company, 93
Martin, 6
McCall Service, 120
Mobil, 8, 66, 76, 77, 106, 117, 118, 122, 124, 126, 129, 147, 187, 193, 219
Mohawk, 70
Montana Grizzly, 87
Moore Oil, 86
Murphy Oil, 7, 153, 158
Murphy-Spur, 7
Mutual Oil, 84

North Star, 6
Nun-Bet-Er, 25

Oklahoma, 136

Pan-Am, 7, 77, 97, 115, 123, 140, 141, 142, 149, 151, 178, 188
Pan-Gas, 21
Parco, 27
Pate, 6, 8, 114, 161
Payless, 124
Peerless, 19
Pennzip, 62
Pennzoil, 38
Peoples Oil Company, 122
Perfect Power, 6
Permex, 108
Phillips 66, 8, 124, 141, 149, 164, 170, 184, 188, 219
Phillips Petroleum, 166, 213
Pow-R, 84
Pure Oil, 7, 85, 86, 91, 93, 94, 100, 113, 119, 122, 128, 133, 138, 139, 145, 183
Pure Pep, 43
Pure, 8, 103, 112, 149, 176, 186, 180, 190, 191, 210

Red Ace, 172
Red Ball, 15
Red Crown, 11
Red Diamond, 198
Red Hat, 5

Red Head, 6, 91
Rex Oil Company, 133, 167, 191, 193
Richfield, 21, 24, 25, 30, 36, 39, 40, 42, 51, 52, 60, 63, 125, 154, 177, 181
Rio Grande, 52
River States, 153
Roosevelt Oil Company, 116

Savings Station, 198
Seaside, 53, 156
Shell, 7, 34, 35, 38, 47, 57, 69, 79, 95, 97, 103, 105, 110, 113, 121, 132, 134, 135, 143, 144, 172, 175, 177, 211
Signal Oil and Gas, 180
Signal, 156
Sinclair, 13, 17, 29, 33, 42, 52, 64, 92, 93, 94, 96, 102, 114, 115, 116, 123, 129, 134, 135, 139, 150, 172, 177, 178, 181, 188, 211
Site, 187
Skelly, 7, 63, 64, 141
Sohio, 33, 44, 45, 89, 125, 126
Speedway 79, 125, 147, 195, 197
Spur Distributing, 153, 202
Spur, 6, 7, 8, 41, 81, 102, 103, 117, 133, 141, 154, 158, 159, 172, 177, 210
Standard Oil of California, 5, 8
Standard Oil of Indiana, 6
Standard Oil of Kentucky, 8
Standard Oil of New Jersey, 6
Standard Oil of Ohio, 31
Standard, 8, 10, 11, 18, 28, 32, 33, 36, 65, 93, 94, 114, 115, 156, 180, 182, 183, 192, 220
Sunoco, 34, 221
Sunset, 46
SuperAmerica, 204, 209

Tankar, 6, 96, 108, 114
Texaco, 7, 9, 11, 14, 18, 19, 49, 75, 81, 92, 93, 95, 97, 99, 100, 102, 111, 119, 132, 134, 138, 147, 166, 175, 176, 183, 189, 192, 206, 221, 222, 223
Texas Oil Company, 5
Tops Oil Company, 201
Trackside, 6
Travelers Oil, 170, 171

Union Oil, 48
Union, 21, 32, 36, 41, 46, 48, 130, 191
Unocal, 210
Urich Oil, 6
Utoco, 8, 115

Vickers, 158, 159, 160

Wake Up, 60, 61
Wareco, 203
Webb Cut Price, 6
White Eagle, 43
White Rose Gasoline, 23, 76, 77, 127
Wilco, 208
Wilshire, 68
Wm Penn, 38

X-L Oil Company, 128, 143

York Oil Company, 71